Nothing

A Philosophical History

ROY SORENSEN

UNIVERSITY PRESS

OXFORD
UNIVERSITY PRESS

Oxford University Press is a department of the University of Oxford. It furthers the University's objective of excellence in research, scholarship, and education by publishing worldwide. Oxford is a registered trade mark of Oxford University Press in the UK and certain other countries.

Published in the United States of America by Oxford University Press
198 Madison Avenue, New York, NY 10016, United States of America.

Library of Congress Control Number: 2021033934
ISBN 978–0–19–974283–7

DOI: 10.1093/oso/9780199742837.001.0001

1 3 5 7 9 8 6 4 2

Printed by LSC Communications, United States of America

To all my daughters.

Even in number,
Just like my sons who number two.
But my daughters are neither two nor four nor six nor eight.
A number that ends in 0 makes some celebrate
They think my mate,
A hero
But rest assured,
My daughters number zero.

Contents

V. DIVINE NOTHING

VI. SCIENTIFIC NOTHING

VII. SECULAR NOTHING

List of Figures

Preface

> In the beginning God created the heaven and the earth. And the earth was without form, and void; and darkness was upon the face of the deep.
>
> —Genesis 1:1–2

Creation stories try to explain how everything originates from nothing. They leave something out. Nothing also has a history. This book aims to tell it.

Books about nothing may go back for billions of years. So say astronomers who conjecture that civilizations formed soon after the universe cooled to form stars and planets. What did the antennas of these historians miss that might be captured in this book?

The hominid side of nothing. I start with a cousin of *Homo sapiens* who picked up a pebble with holes that seemed to make faces (fig. P.1). Many faces later (each chapter pairs a philosopher with an absence), I conclude with Bertrand Russell's precise analysis of how 'Caspar does not exist' could be true (chapter 22).

About the fifth century BC, three civilizations independently and simultaneously began to philosophize about nothing: China (chapter 3), India (chapters 4 and 5), and Greece (chapters 6–10). Their luminaries had previously focused on what is the case. Search beams of consciousness swept majestically across the field of being. But then there was a black-out. Voices from the dark spoke now about what is *not* the case.

Behold! The holes in a sponge are absences of sponge! Holes are what make the sponge useful for absorbing liquid. The sponge can exist without the holes. But the holes cannot "exist" without the sponge. Holes are parasites that depend on their host. Yet the two get along well. Without holes, there would not be so many sponges in your home!.

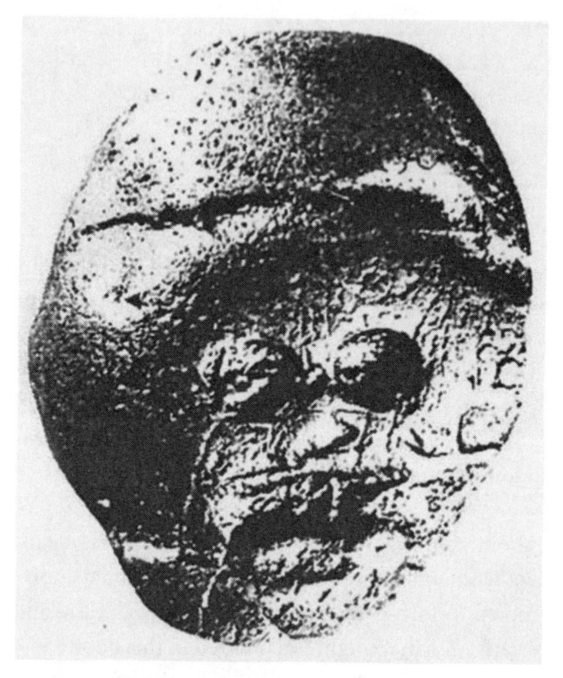

Figure P.1 Photograph of Makapansgat Pebble by Patrick Nagel and Raymond Dart

Your house is made more cozy with doors and windows. Yet these amenities are metaphysical amphibians. When you stand under a door frame are you inside or outside? Is the door an opening? Or is the door a rectangular plug mounted on a hinge? We experience the same indecision when trying to decide whether windows and skylights are absences or presences.

You see the tiny boundary of the dot that ends this sentence. Is that boundary black or white? Not black, because anything that is black is part of the dot. Not white, because anything white is part of the dot's environment.

Boundaries are parasites of material hosts. But they are also parasites of immaterial hosts such as your shadow. Your shadow is a hole *you* bore into the light. Your shadow depends on both you and the light. You and light are each mysterious. Your shadow partakes of both mysteries.

Being is riddled with nonbeings. Why were the riddles first posed 2,600 years ago? Why all at once? This negative turn in world philosophy is the coincidence that inspired me to write *Nothing: A Philosophical History*.

My goal was to find a common factor that could explain the simultaneous and independent shift in perspective. My best candidate is a copying trick.[1] Any experience of an event can also be explained by the parasitical hypothesis that the event was merely dreamt. The parasite copies the consequences of 'The event was perceived'. Consider a little girl who is awakened by sounds of her parents. Embarrassed, they assure their daughter that the scene is a dream. In the morning, her parents keep up the lie. Their teamwork overrides the child's perception. The parasite hypothesis converts the daughter. After she wises up, the daughter concludes that *any* waking experience can be re-explained as a dream experience. She generalizes: *all* waking experiences can be explained as a dream. Dream skepticism is kept a live option because she dreams every night.

Host hypotheses have defenses against the insinuation that they are falsehoods that merely have true consequences. Gilbert Harman (1988, 40) notes that the host has the advantage of simplicity. The parasitical alternative requires editing the original tale told by the host. A further advantage for the host is our preference for hypotheses that *predict* truths rather than merely *accommodate* host's discoveries. Any parasite can wait for a host theory to make the discovery and then fudge its calculation to match the host's prediction. To stand out from the competition with other parasites, the parasite needs to generate its own predictions. In the 1960s, linguists knew that some phase structure grammars could co-opt the predictions of transformational grammars. But the linguists only seriously subscribed to phase structure grammars after phase structure grammars made their own predictions (Harman 1963).

The parasite has its own simplicity. Parasites are not lumbered by the commitments of their host. Old men in ancient Rome reported

[1] A few people report never dreaming. Nevertheless, there are many dreamers in every culture. "The father of history" Herodotus disagrees. In book IV of Histories he reports that many thousands of years ago in North Africa, the Atlantans never dreamt.

dark specks floating in their eyeballs. Skeptics said the old men were hallucinating the specks. The hallucination hypothesis avoids commitment to specks. It also predicts that no specks will ever be discovered by future observers of old eyeballs. The parasite lost that bet! But at least the parasite was making some novel predictions.

All parasitical theories postulate psychological mechanisms that explain how perceptions could occur in the absence of the represented events. During the late Warring States period (476–221 BC) Master Zhuang dreams he is a butterfly (fig. P.2). But then the sage wonders whether he is a butterfly dreaming he is a man. That is lovely as poetry. But butterflies lack the neural and social infrastructure to dream they are men. If you dream you are a man, you might be woman. You might

Figure P.2 *The Butterfly Dream*, by Chinese painter Lu Zhi (ca. 1550)

be a child older than two—the age at which dreams acquire plot lines. But you are no butterfly.

Typically, a parasitical theory depends on their host's survival. This means the host could later throw off the parasite. For instance, atomism was long parasitized by the hypothesis that the doctrine is useful make-believe. Some of the subtlest improvements of atomism were undertaken by physicists who simply wanted make the most from the host. In the beginning of the twentieth century, Albert Einstein interpreted the random motion of pollen grains in water as collisions with atoms. This explanation required physical atoms rather than make-believe atoms. The parasite was expelled. The host retained the improvements engineered by the parasite.

Parasitical improvements of hosts are now deliberately cultivated. In null-hypothesis methodology, scientists are required to entertain a rival hypothesis that an apparent cause is a mirage of correlation.

A parasite can artificially prolong the life of its host. When parasites castrate crabs, the crabs get into fewer fights. Geocentric astronomy would have been killed off by heliocentric astronomy. Surveyors intervened. They have guaranteed geocentric astronomy a long future as a falsehood that simplifies measurement. Newton's physics survives as a limiting case of Einstein's physics. A magnanimous victor will live in harmony with neutered adversaries. This prospect leads the old guard to nervously cross their legs when the young begin praising their elder's theories as limiting cases of a fresh theory.

Contemporary parasites owe their sophistication to the private reality opened by Christian introspection (chapter 13). Previous thinkers had treated the utterer of 'It seems to me that Jesus wept' as refraining from reporting anything. Saint Augustine treats the sentence as a report of a mental fact. This inner fact can only be directly observed by the speaker. He has privileged access to matters that others can only infer from his behavior.

These subjective threads were woven into a universal coordinate system by René Descartes. As the thread count increased, the Cartesian veil of ideas eventually enveloped the material world. And then, in a great vanishing act, George Berkeley pulled away the matter beneath the veil. The veil remained afloat. Where did the external world go?

The magician answered, "To be is to be perceived. There was never any external world to begin with."

Berkeley's immaterialism could only thrive on a rich diet of materialist hosts. Thanks to political scale, China and India achieved a critical mass of hosts about 500 BC. Lightly populated Greece achieved critical mass by commerce rather than by indigenous fecundity. The intricate coastline and islands placed these scattered people at a crossroads between civilizations. Their large neighbor civilizations had come close to achieving the critical mass needed for parasites to bore their holes. The Greeks accumulated these near misses into a hit. After Aristotle's student Alexander invaded India, the Greeks were able to import some of the parasitical breakthroughs pioneered by the Hindus and Buddhists. By copying the Indian copycats, the Greeks conquered the whole world at an intellectual level.

In Europe, this conquest was initially hampered by Christianity—but eventually helped. I focus on the Christians because of their love-hate relationship with atomism. Abhorrence of the vacuum had been the default attitude in the West since Parmenides. But in deference to the Genesis 1:1-2 scripture quoted in the epigraph, medieval philosophers such as Thomas Bradwardine were able to make a safe space for the void (chapter 16). This was the space later exploited by Isaac Newton's physics (chapter 17).

When I gaze outside my eastern window, I see Elon Musk's SpaceX rockets launching toward the sun. The drama of rocket after rocket is otherworldly.[2] I had never seen a rocket before moving to Austin, Texas. Now I view launches as a weekly spectacle. Musk himself is a spectacle. He espouses the simulation hypothesis: almost all consciousness is the effect of computer programs. Instead of being at the Boca Chica, Texas, launch site in the year 2021, Elon Musk exists far in

[2] Presently, Mr. Musk is better at launching than landing. When reusable rockets were first proposed in the 1960s, the mathematical physicist Stanislaw M. Ulam had technical reservations. Instead of boring air force generals with his pessimistic calculations, Ulam slipped into a meeting led by an industry promoter of the rocket recyclers. During the shuffle of plots and graphs, Ulam whispered to General Avner Gardner, "This sounds to me like a proposal to use the same condom twice." The general burst out laughing. The phallic comparison was repeated in whispers around the table. "Perhaps the joke saved the United States some millions of dollars in expenditure for what would have been a pointless and impractical work at the time" (Ulam 1976, 256).

the future. He has no hands, no heart, no head. Mr. Musk is an invention of future historians studying their ancestors who lived way back in 2021. The historians compare what actually happened in 2021 to what would have happened if there are had been an entrepreneur pioneering internet banking, electric cars, and a mission to Mars.

If I had lived a life as improbable as Elon Musk's, then I might be tempted to think it all a dream. But Nick Bostrom's (2003) original support for his simulation hypothesis is a statistical argument that does not depend on extraordinary events. According to Bostrom, ordinary people ought to assign a surprisingly high probability to the simulation hypothesis. Bostrom has persuaded the eminent philosopher David Chalmers (2022, 100) to assign a probability of at least .25 to the simulation hypothesis. The astrophysicist Neil deGrasse Tyson assigns a probability greater than .5. The biologist Richard Dawkins, equally proud of his immunity to philosophy, takes the simulation hypothesis seriously.

Sound familiar? The simulation hypothesis is the latest parasitical theory. This twenty-first-century specimen incorporates the era's technical and social novelties. A live option! Whereas skeptical uses of parasites completely empty out the past and the material world, the simulation hypothesis innovates by preserving the past. That is why Dawkins, an outspoken evolutionist, can allow that the simulation hypothesis is true. The simulation hypothesis entails that Dawkins' heavily historical scientific theory is true. What gets deactualized is the *present*!

I would be disappointed to learn my recent experiences do not correspond to reality. *Alas, I never saw a rocket! 'Twas but a bit of coding.* Others find their disembodied alternative as enchanting as the ancient Hindus who conceived of themselves as dreamed by the gods.

Acknowledgments

"We'll get back to you," said I, late on a deadline. An impatient editor replied, "There are three classes of people who use 'we' instead of 'I': Royalty, schizophrenics, and men with tapeworms. Which are you?"

None of the above! My 'we' is constituted by the people behind this million-year history of nothing. David and Stephanie Lewis started me on a slippery slope to the Void with their charming dialogue "Holes." One of their fans, Achille Varzi, gave me a tour of pavement cracks, cavities, and tunnels in Manhattan. His coauthor of *Holes and Other Superficialities*, Roberto Casati, introduced me to the shadows of Paris. Roberto's highly visual chronicle *The Shadow Club* awakened my interest in the history of absences—and in their characteristic appearances. Edward Grant's medieval history of the vacuum taught me what science and religion have in common: *nothing!* Ironically, the Void of the ancient atomists became crucial to the very Christians who spent a thousand years burying atomists as atheists. Grant's subtle connections between the scholastics and Isaac Newton helped me relate objective nothingness with its subjective side.

The subjective side is fed by a stream of comical encounters between philosophers and nothing. Not for nothing was Democritus known as the laughing philosopher. Atoms and vacua make for odd bedfellows. His critics tried to laugh him down by pointing out the incongruous relationship between these opposites. Cheerful Democritus took the jokes as revealing our absurd expectations. He inaugurated a tradition of passing on philosophical jokes to students. And I am happy to build on this lore. I thank the many philosophers and historians who have treated me as a repository for this scholarly humor. Recent donors to the repository include James Dee, Matt Evans, John Heil, Kathleen Higgins, Stephen Read, and Stephen White.

I am grateful to the University of Texas at Austin for providing a College Research Fellowship for 2020–2021. This has been in addition to the research funding furnished by a Darrell K Royal Regent

Professorship in Ethics and American Society. As a Professorial Fellow at the University of St. Andrews in Scotland, I have also benefited from an institution founded in 1413.

A few of the chapters incorporate material from past essays. Chapter 1 incorporates opening passages of "The Vanishing Point: Modeling the Self as an Absence," *The Monist* 90/3 (2007): 432–456. Chapter 11 incorporates portions of "The Symmetry Argument" from the *Oxford Handbook of Philosophy of Death*, edited by Ben Bradley, Fred Feldman, and Jens Johnansson (Oxford University Press, 2013). Chapter 15 contains material from "Parsimony for Empty Space," *Australasian Journal of Philosophy* 92/2 (2014): 215–230. The chapters on Bergson and Sartre incorporate sections of "Perceiving Nothings," which appeared in *The Oxford Handbook of Philosophy of Perception*, edited by Mohan Matthen (Oxford University Press, 2015). The chapter on Bertrand Russell incorporates some paragraphs of "Unicorn Atheism," *Nous* 52/2 (2018): 373–388. I thank the editors and journals for permission to use this material.

During the Covid-19 pandemic in 2020 and 2021, I could not enter the Austin Nature and Science Museum to photograph their human sundial (presented by the City of Austin and MathHappens). I received inside help from the exhibit coordinator, Joshua Random. He photographed his son Alpha Random in the role of gnomon (figure 14.1).

I also thank my son Maxwell Sorensen for figure 18.5 and his brother Zachary Sorensen for figures 8.4 and 19.2. I thank my wife, Julia Driver, for them, and for many years of a philosophical marriage.

Finally, I thank the editor, the copyeditor, and fact-checking students for removing falsehoods with the efficiency of my automatic spellchecker. As for the falsehoods that have survived, I hereby retract them all. Enjoy reading the first history of philosophy free of any false assertions!

Nothing

I

NOTHING REPRESENTED

1

The Makapansgat Hominid

Picturing Absence

Three million years ago, a hominid found a pitted, waterworn pebble. You can find it in the preface (figure 0.1). The holes make the pebble resemble a face—indeed different faces when viewed from different orientations. When viewed upside down, the face resembles a toothless old man. Stare at the concave eyeholes. The eyeholes pop out.

This gestalt switch to a bug-eyed man may have been experienced as a power of the pebble itself (rather than an effect of a visual system vacillating between two interpretations of the stimulus). The magical pebble was carried a long distance to a cave in Makapansgat, South Africa. This cave has the remains of *Australopithecus africanus*.

Under one theory of artistic progression, the first stage of "found art" triggers a second interventionist stage. The proto-artist sees an object as a near miss of something worth perceiving for its own sake. "If that bump were absent, the rock would resemble a face." By subtracting the bump, passive appreciation graduates to artistic intervention. This intervention educates the proto-artist into then discerning near misses of near misses: "If that big bump were absent, the rock would look like the rock that was a near miss of a rock that looked like a man." Apprentice artists lengthen the chain of near misses passed down from their masters. Countless counterfactuals later, Florentine sculptors saw a giant block of white marble as encasing the figure of David. But was David lying down as the block lay? Where was the head of the slain Goliath? Michelangelo upended the roughly begun block, omitted Goliath, and portrayed David prior to the fight, leaving a rock hidden in his hand.

Neanderthals graduated beyond subtraction. They drew geometrical images such as lines, ladder-like configurations, hashtags. Instead of chiseling out something that already existed, they made something new. If the Neanderthal hashtags were depictions of palm creases, then they also took the step of creating representational art.

Many absences have a characteristic appearance. The total darkness of a cave looks uniformly *black*. Shadows resemble the objects that cast them.

Shadows are holes in the light. Ironically, they can be painted but light cannot. The landscape water colorist J. M. W. Turner gave the *illusion* of painting light by painting shadows. Light emerges by contrast with darkness. When a critic complained that Turner had painted the sails as black as he could, Turner replied, "If I could find anything blacker than black I'd use it."

Relative absences are depicted by omitting expected features from a host. Absolute absences are more challenging. Nevertheless, absolute absences often have a characteristic appearance. Since total darkness looks black, a totally black canvas more naturally depicts darkness than a totally red canvas.

After Filippo Brunelleschi (1377–1446) introduced linear perspective to render the recession of space, artists used the vanishing point to symbolize mysteries. One of the mysteries is whether it can be seen. Consider how rails seem to converge in the distance (fig. 1.1). This point on the horizon at which receding parallel lines meet is the vanishing point. I can see that it is at the center of my visual field. But can I see the vanishing point?

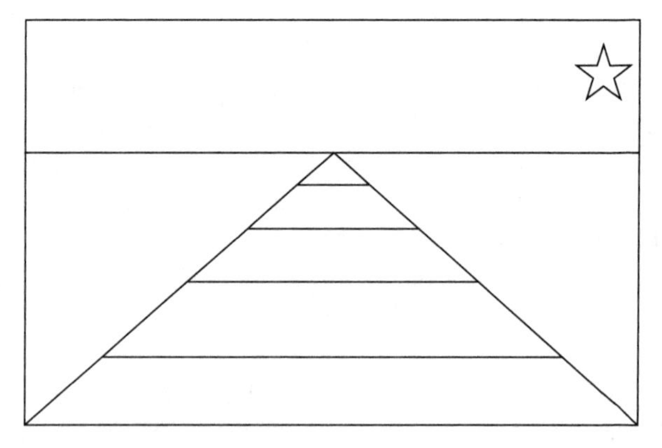

Figure 1.1 The vanishing point

On the one hand, the vanishing point seems visible because I can point straight at it. I can move my thumb over the vanishing point and thereby block my view of it.

There is no special problem with the vanishing point being a *point*. I see the points at which the railroad ties meet the rails. And they lead right up to the vanishing point.

A point is a feature rather than an object in its own right. The point of a dagger is not a part of the dagger in the way the blade and handle are parts. One might assimilate the perception of points to the perception of corners. A corner is always a corner of some host object. One sees the point at which three sides meet by looking at a cube. By recognizing the parasitical nature of points, we avoid the fallacy of reification.[1]

On the other hand, I see something only if it looks a particular way. The vanishing point does not look like anything. It lacks a color or shape. In Leonardo da Vinci's *The Last Supper* the vanishing point coincides with the right eye of Jesus. But the vanishing point does not look like an eye!

If observers were in the scene, they could bend over a point at which a rail is attached to a wooden tie. Yet an observer cannot bend over the vanishing point and truthfully say, "There is the vanishing point." However far the observer journeys, he will not reach the vanishing point.

The vanishing point is like the edge of my visual field. I can locate this boundary by attending to what I see. But I cannot see the edge of my visual field. When I peer at a peephole at a construction site, I can see both the edge of the peephole and the scene through the hole. But normal vision is not like looking at a peephole. The edge of my visual field is an outer limit. The vanishing point is an inner limit.

I see things within my visual field but not the visual field itself. If my visual field were an object of vision, then there would be a second field

[1] The showman P. T. Barnum exploited this presumption that words refer to objects (*res* in Latin). To prevent customers from lingering at his exhibit of exotic animals, he posted a sign over a door: "To the Egress." The door would lock behind the curious customer—who found himself on the street.

of vision encompassing the first. If each visual field were itself an object of vision, then there would be an infinite hierarchy of visual fields. I would see something akin to the embedded scene produced by a pair of mirrors that reflect each other's reflections.

The simplicity and directness of the visual field make it perfectly "transparent." I can see a lightly fogged windshield while simultaneously seeing through it to the road ahead. I cannot see through my visual field in the same way. In "The Refutation of Idealism" G. E. Moore (1903) analyzed his consciousness of blue. He found something corresponding to blue but nothing corresponding to consciousness. This makes consciousness seem empty. According to Moore, materialists react by denying the existence of consciousness. In *Being and Nothingness*, Jean Paul Sartre affirms the existence of consciousness but characterizes it as nothingness. Moore would say that Sartre is mistaking an absence of representation as a representation of absence. I see through my spectacles but am not seeing an absence of spectacles.

The vanishing point has a central position in picture composition. Accordingly, Christian artists marked the mystery of the Eucharist by placing the transubstantiation at the vanishing point.

The technique is adapted for secular purposes by Joseph Wright of Derby in his 1768 *An Experiment on a Bird in the Air Pump*. Wright places a vacuum at the vanishing point (fig. 1.2). Wright's use of a white bird rather than the more usual mouse may reflect the Christian use of the dove as a symbol of Holy Spirit on the Eucharistic vessel. The air of a black mass is conveyed by the gothic lighting and the full moon through the window (though scientific demonstrations were indeed scheduled during full moons for the sake of safe travel in the dark).

In Wright's era, the void was commonly pictured as completely empty. Experimentalists claimed to have cut through millennia of skepticism about the void by creating a vacuum (reversing God's creation of matter from void). Skeptics objected that there must be something inside the "empty" glass; how else could light be transmitted through it? In *The Leviathan and the Air Pump* Steven Shapin and Simon Schaffer argue that the vacuum pump was more creative than

Figure 1.2 Joseph Wright, *An Experiment on a Bird in the Air Pump*

scientists care to admit. They contend the vacuum was a social creation, not a pure observation.

This book is a general history of nothing, not a special study of the vacuum. But on my interpretation, absences are discovered rather than invented. That is the best way to make sense of what scientists say about holes, cracks, shadows, darkness, and silence.

Admittedly, scientists editorialize beyond their expertise. When asked which hole develops first in human beings, an embryologist responded that we are deuterostomes, meaning "mouth second." The first hole to form is the anus. The embryologist concluded that we all begin as assholes.

The perception of absences predates human beings. The earliest "ears" heard silence. The earliest "eyes" saw little shadows grow into big shadows, and big shadows become total darkness.

Animal awareness of absences bedevils behaviorists. According to the law of effect, each response is caused by a stimulus. Yet birds react to the absence of light with nervous excitement. Infant monkeys respond to the absence of a mother with despondency. Captive animals respond to the general absence of stimulation by pacing and other repetitive behavior. A zoo is a zoo of boredom.

Animals take advantage of absences such as darkness and holes. Caves became contested resources. The Chauvet cave in Lascaux, France, contains tracks of many animals. The cave also has the earliest depictions of absences (fig. 1.3).

The European male lion, now extinct, is depicted as lacking a mane. Since manes do not fossilize, biologists have cited this painting as evidence of manelessness.

How do the biologists know that the maneless lions are male? By another absence. Female lions are depicted without scrotums.

The Chauvet depictions of absences are sophisticated. Butchered animals are represented as limbless. But close-ups of other animals do not depict them as limbless or as only having the side facing the viewing or as only having a surface. Chauvet artists expect the viewer to distinguish between representing an absence and an absence of representation.

The depictions are also dynamic. They show animals in motion. We interpret the double-drawn head as moving, changing over time from what is to what will be.

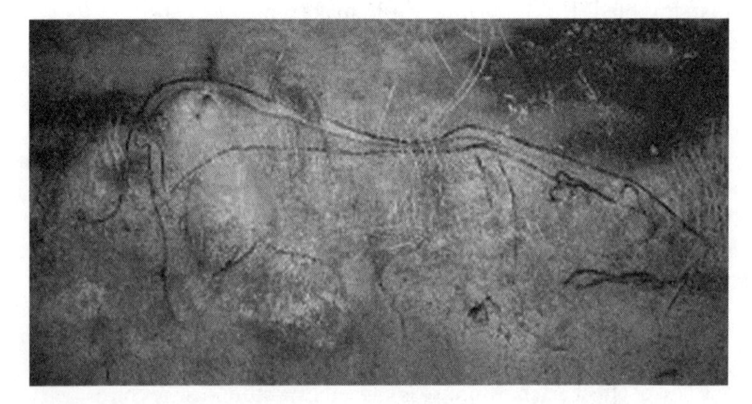

Figure 1.3 Two black lions. Bradshaw Foundation Cave Art Gallery

Painters may have exploited the flickering lights of grease lamps to foster the illusion of motion. Grease spots in the caves suggest that the paintings were staged. Shadows may have been deliberately projected onto the paintings. Sounds and echoes may have further animated the figures. Placing candles behind the viewers would have introduced an element of audience participation. The viewer's shadow would have become a dynamic part of the scene.

If you interpret the movements of your shadow as *actions* rather than effects of your actions, you have a *ghost*. Dreams reinforce this attribution of beliefs and desires to immaterial beings. The painting in figure 1.4 is a 20,000-year-old depiction of a man, a bird, a broken spear, and a disemboweled bison. The scene has been subject to much interpretation. A hunting disaster? A magic spell? According to the neuroscientist Michel Jouvet, the phallus is the key that unlocks the mystery; the man is dreaming. Jouvet pioneered research showing that dreams occur in a paradoxical phase of sleep in which the brain is almost as active as the waking state. A subtle sign of this activity is rapid eye movement below the eyelids. But in most male animals there is also

Figure 1.4 Bird man of Lascaux. Attribution Peter80

the marker of tumescence (which occurs independently of whether the dream is erotic). This connection between somnolent erections and dreaming would have been salient in open Paleolithic sleeping conditions. The bird is a metaphor for the soul across diverse cultures. The dreamer's mind flies to distant places and times. Thanks to dreams, we are natural-born dualists. We distinguish the mind from the body because the mind can be separated from the body, at least in dreams, and perhaps at death.

If I existed after death (or before birth), the fact that I exist during the present period would be less surprising (Huemer 2021). A being who spent most of his time existing would have a much better chance finding himself alive in 2021. A being who lasts less than century would have to regard himself as the winner of a lottery with infinitely many tickets. Even an atheist may find himself raising the probability of reincarnation or an infinite afterlife.

Many pictures became symbolic as interpreters coordinated. Some of these standardized pictures become pictograms. They could be sketched schematically without loss of meaning. When organized syntactically, pictograms substitute for speech.

Only human systems of communication have a symbol for negation. A lion can reject food, but a lion cannot *deny* a proposition, attribute *falsehood*, or be confused by contradictory assertions. Lions never lie.

Music is also unique to human beings and deeply universal among them. Paleolithic flutes have a pentatonic scale. This scale is culturally common because of self-tuning. Flutes with evenly spaced holes are out of tune with themselves. Ancient people must have patiently experimented to eliminate dissonance. The spacing of the absences allows archeologists to play "The Sound of Silence" on a 30,000-year-old flute.

Instead of simply burying corpses, cave people included musical instruments, tools, and jewelry in their burials. They appear to have hoped for an afterlife. Perhaps human beings relate to their bodies as their bodies relate to shadows. This spiritual simile makes sense only in contrast to a starker alternative in which death is the cessation of existence. We naturally opt for the stark alternative for animals. An ibex is just the living body of an ibex. A dead ibex is an ex-ibex.

Given our resemblance to other animals and the implausibility of crediting animals with any afterlife, the hope of human afterlife can only be sustained by rupturing the continuity between human beings and animals. Fear of permanent annihilation led human beings to improvise myths that forge a protective gap between people and other animals. In the next chapter, we turn to a civilization that standardized some of these myths.

2

Hermes Trismegistus

Writing Out Absences

The earliest writing developed from the need to seal containers. The barrier created the problem of remembering the contents of the container. Since the seals were made of clay, proto-clerks impressed a mark indicating the amount and nature of the contents. Even empty containers needed to be sealed to prevent contamination. So there were marks for empty receptacles. This outside-in origin of writing was reversed by Victorian potters who inscribed the letter riddle OICURMT *inside* the bottom of their receptacles.

A big riddle: What is the largest possible receptacle? Philosophical answer: the universe. The container model of the universe puts us in a position to ask a question that no Egyptian asked: Why is there something in the universe rather than nothing?

You are inside the largest container. Consequently, you know there is something rather than nothing. Particular things within the container can explain other particular things within the container. Your parents explain your presence. Your grandparents explain the presence of your parents. But none of the particular things can in the universe explain why there exists at least one particular thing rather than no particular things in the universe.

Once the empty world is accepted as possible, it seems simpler than any occupied world. For the presence of an object requires an explanation, not its absence. "Who ordered that?" asked the Nobel laureate Isaac Rabi when the muon was served up by experimental physicists in 1936. None of the contending theorists expected this heavy cousin of the electron.

After a couple of centuries, Mesopotamian clerks realized that the marks on the containers could be made on portable tablets rather than on the seals of the containers. These records could be stored off-site.

This convenience made the descriptions more liable to be false. The reader had lost easy access to the original items under discussion.

The potential for misrepresentation grew with the spread of writing. The more that could be said, the more that could be said falsely. Combinatorial systems of representation, such as the binary numeral system, represent infinitely many possibilities with a small stock of symbols and rules for combination. All natural languages work in this recursive fashion. An alphabet allows writing to match the expressive power of speech.

The Egyptians developed the first alphabet about 3100 BC. The telepathy of silently communicating with absent agents led the alphabet to be regarded as an invention by Thoth, scribe of the gods (fig. 2.1).

Egyptian gods believe what they read. Temple writers try to make their claims true by exploiting the remoteness of their overlords. Pharaohs chiseled out the names of those they wished to consign to oblivion . . . and chiseled in their own names, grabbing credit for the monuments erected by those they erased.

The ancient Egyptians were prolific propagandists. Military historians consulting the ancient records are impressed by the uniform success of Egyptian armies. Even when the empire declined, the Egyptian armies were always winning—just closer to home.

Much Egyptian writing survives because prayers, spells, other genres of word magic were stored in tombs. The topic of death makes this spiritual library appear philosophical. But this officious papyrus work was no more theoretical than the paperwork presented to contemporary gatekeepers of the Arab Republic of Egypt.

I agree with those Egyptologists who require philosophy to be based on arguments open to public inspection.[1] There are philosophical writings attributed to Hermes Trismegistus, an ancient Egyptian philosopher who systematized the religion of his day in a collection of writings called the *Hermetica*. In addition to philosophy, there are alchemy, potions, and magical spells. The existence of Hermes Trismegistus would fill a gap felt by those who believe

[1] I disagree with Immanuel Kant, who in "The End of All Things" further requires examination of the presuppositions of moral agency, such as freedom of the will. Kant concludes there was no philosophy in the Orient.

Figure 2.1 Thoout, Thoth Deux fois Grand, le Second Hermès. N372.2A, Brooklyn Museum, Attribution Jean-François Champollion

that the highly civilized Egyptians must have had a philosopher to head the arms and legs of government, architecture, and medicine. An absence of indigenous philosophers would leave Egypt an acephalic giant.

The quest for a head explains Hermes Trismegistus's ability to survive a string of scholarly decapitations. Following each refutation of his existence, Hermes resurrects. After disappearing into the Dark Ages, Hermes Trismegistus reappears in late medieval Europe—only to be dunked back into nonbeing by the scholastics. Once the scholastic doubters lost influence, Hermes resurfaced during the Renaissance. He continues to intrigue contemporary New Age thinkers—despite stylistic analysis that dates the *Hermetica* to AD 300.

Undocumented philosophers are on the wrong side of history. Historians write them out of chronicles. These scholars substitute better-credentialed figures. The accidents of document preservation play a disproportional role in peopling the past.

Hermes Trismegistus is a rugged survivor of this publication bias. I do not begrudge him a role in history. Indeed, in this book I counter the historians' prejudice against nonexistent philosophers. When other authors learn that a philosopher may have never existed, they tear his page from the history of philosophy. My reaction is to lightly shade in the subsequent page. The name resurrects through my scribbles. My reverse discrimination reveals how much philosophy enters through wishful thinking, slips of the tongues, and pranks.

In "Speaking of Nothing" Keith Donnellan (1974) flips denials of existence into positive postulations of "referential blocks." A name of a planet is like a perception of a planet. Just as the planet explains the wandering white dot in the astronomer's visual field, the planet also explains why he exclaims, 'Neptune!'. Reference succeeds when the planet caused the representation in the appropriate way.

Occasionally, this normal causal process is disrupted. In astronomy, utterances of 'Vulcan' originate from a misperception or a mishearing or a lie. When critics discovered there was a glitch for 'Vulcan' (which was designed to refer the planet responsible for the perturbations of Mercury), they uttered the negative existential sentence: 'Vulcan does not exist'. The sentence *seems* to be talking about what does not exist. But Donnellan says it is really talking about what does exist: a monkey wrench somewhere in the mechanism of naming. 'Vulcan does not exist' declares a disruption of the causal chain that normally connects present usage of a name with past contact with bearer.

Simon of Dacia is described as the author of the Corpus Philosophorum Danicorum Medii Aevi III. But the name is a fabrication of Amplonius. This bibliophile listed books by authors. When Amplonius did not know a book's author, he filled the gap with a dummy name. Donnellan translates 'Simon of Dacia does not exist' as: *There is some referential block that caused 'Simon of Dacia' to be uttered.*

Referential blocks are typically mistakes rather than lies. For instance, scholars inadvertently cite "articles" that are merely samples of proper citation style. The following phantom article has been cited over a thousand times:

> Van der Geer, J., Hanraads, J.A.J., Lupton, R.A., 2000. The art of writing a scientific article. J Sci. Commun. 163 (2) 51–59.

Neither the article nor the journal exists. Busy authors either cite the article without reading or inadvertently leave in the template. Copycat citations commence. Since frequent citation is a mark of merit, the citations snowball.

Snowball citations would have been evaporated had scholars applied the hellfire of Katherine Frost Bruner: "Incidentally, a sin one more degree heinous than an incomplete reference is an inaccurate reference; the former will be caught by the editor or the printer, whereas the latter will stand in print as an annoyance to future investigators and a monument to the writer's carelessness" (1942, 68). Editors of journals concurred with Bruner's ranking of inaccurate quotation as even worse than incomplete quotation. They quote her in their style guidelines. Authors third the editor's seconding. Through such seconding, thirding, and fourthing, the last part of Bruner's sentence has been repeated over ten million times in nine different languages. "The irony is that the vast majority of those authors who have been inspired by Katherine Frost Bruner's message about the importance of complete and accurate references have not managed to quote her accurately" (Rekdal 2014, 746). The most common alteration of Bruner's message is to subtract the assurance that the inaccuracy will be caught by the editor. The editors' incomplete quotation has the effect of shifting all of the responsibility of quotation to the author.

Who is responsible for philosophical problems? Philosophers bear some responsibility. But it is popular to shift all the responsibility to their shoulders by accusing them of creating the problems they try to solve. Outsiders picture philosophers as dogs chained to trees. In the morning the dogs wander around their trees in smaller and smaller circuits. As the length of their leash dwindles, the dogs bark louder and louder. At lunch time, caretakers return to untangle the hoarse dogs. In the afternoon, the liberated dogs return to their work of creating problems for themselves. Philosophers who present themselves as friends of common sense concur. George Berkeley chides his colleagues: "Upon the whole, I am inclined to think that the far greater part, if not all, of those difficulties which have hitherto amused philosophers, and blocked up the way to knowledge, are entirely owing to ourselves—that we have first raised a dust and then complain we cannot see" (1948, 2:26). Could problems of nothingness have originated with philosophers?

The absence of Egyptian philosophers shows that the dread of nothingness was not started by philosophers. Philosophy is instead an *effect* of that fear of oblivion. In the twentieth century, Ludwig Wittgenstein tries to cure us of philosophy. Ancient Egypt was free of any such neuroses for millennia. Civilization has been free of philosophy longer than it has been bound by philosophy.

Egyptian religion was practical. Heaven was envisaged as an upgrade of ordinary life. Everything is better by degree: finer garments, more servants, milder weather. One qualitative difference can be discerned from Egyptian pictures of heaven: an absence of graveyards.

Since the Egyptians did not know the details of what would be needed for the afterlife, they packed for everything. There are tools, food, and cosmetics. Much of the cache was miniaturized—including servants. Magical spells would bring Tutankhamen's mummified ducks back to life (so that they could be reslaughtered in the afterlife). The spells were recorded in manuals. An Egyptian tomb resembles the Apollo 11 lunar lander.

Also in the tomb are incidental writings left by the builders and custodians of the tomb. The reading material is varied: mathematics, medicine, riddles, and fiction. The mathematics is based on secret recipes. There is no tradition of proof.

The closest approximation to philosophy in ancient Egypt is wisdom literature. One features argument and counterargument: "The Dispute between a Man and His Ba" from the Middle Kingdom (2008–1630 BCE). The ba is the part of the soul bearing personality. It is represented by a bird that flits above one's head. Another part of the soul, the ka, is a body double (akin to the mummified body that will later be rejuvenated). The record of this dialogue is incomplete. Ancient scrolls decay from the outermost page inward and from the innermost page outward. (The first page serves as the outer wrapping. It gets battered. The last page is the most tightly wound. It is under the greatest structural stress.) This systematic obscurity of beginnings and endings is an apt metaphor for history.

For whatever reasons detailed in the missing beginning of "The Dispute between a Man and His Ba," the man finds life tiresome. The man speaks longingly of death. His ba opposes suicide. The despondent man offers to build a mortuary temple and have his descendants make offerings. But his ba retorts that he will not cooperate with a suicide's resurrection. This threat frightens the man. If he is not resurrected, then he faces a future of infinite nothingness.

Oblivion so terrified the Egyptians that their religion has no need for hell. There is only heaven and nothingness. Alarmed, the man warns that his extinction would also go badly for his ba. But the ba continues to shame him for insufficient fortitude. Disgusted, the ba leaves. Abandoned, the dispirited man half-heartedly continues in forlorn monologue . . . The end?

"The Dispute between a Man and His Ba" owes much of its philosophical appearance to its topic question: Can suicide be prudent? In *The Myth of Sisyphus*, the French existentialist Albert Camus writes, "There is only one really serious philosophical question, and that is suicide" (1955, 3).

The decision to shorten your life resembles the decision to shorten your presence in your homeland. We are all compelled to emigrate to the Unknown. Instead of waiting, the suicide departs before the deadline.

The comparison between suicide and flight to other lands runs into a logical difficulty. Nonexistence is not akin to the Netherlands. Suicide cannot be an escape because there is no "there" to which one flees.

When you came into existence, you did not develop from a nonexistent entity into an existent entity. Nonexistent people do not exist! Therefore, beginning to exist was not a step up from a low level of existence to a higher level of existence. Nor could the end of existence be a decline to a lesser existence. Being a little existent is like being a little identical.

In sum, rational suicide seems to require an impossible comparison between existing and not existing. To choose, one must compare how well you would be in one situation to how well you would *be* in another situation. All prudential comparisons presuppose you exist. That is what the existentialists were barking about.

Notice that the man conversing with his ba never considers objections to the emigration model of suicide. The weary Egyptian does not want to commit suicide in the sense of ending his existence. He just wants to travel to heaven earlier, as measured by how much future terrestrial experience remains.

The man's ba objects that a premature emigration would lower his chance of admission. Prospective immigrants are judged partly by how well they discharged their obligations to their family, kin, and leaders. Cowards and shirkers have their hearts thrown away and are consigned to oblivion.

Early emigration is a momentous decision. But it is not thereby a philosophical decision. Whereas mythology requires a message and religion requires belief, philosophy requires validity. Suicide is puzzling because the conclusion 'I am better off not existing' does not follow from the premise that my existence is bad. Bad is often my best option. Non-existence is never an option. So non-existence cannot outrank existence.

Nor can non-existence under-rank existence. I am glad to exist. But this cannot be because non-existence is a state in which I am worse off. I feel like the winner of a lottery in which the vast majority are losers who never enjoy a sunrise. Maybe the ancient Egyptians felt lucky the same way.

The weary man in the dialogue is not conceiving suicide as an answer to Hamlet's question, "To be or not to be?" Like other Egyptians, the weary man takes it for granted that nonexistence is the worst possibility and thus something that could not be prudently chosen.

Some translators doubt that "The Dispute between a Man and His Ba" really concerns suicide. The only way to extract philosophy from Egyptian texts is through elaborate interpretation. Skeptical Egyptologists regard this exegesis as an anachronistic projection of philosophy into an aphilosophical culture.

Admittedly, ancient Greeks trace their culture to the far more ancient Egyptians. But these status-hungry newcomers do not provide any doctrinal details. Greek philosophers traveled to Egypt as wide-eyed tourists. They saw impressive monuments and practices. But there is a difference between being inspired and being taught. Egypt is an awesome civilization, and philosophy begins in wonder. But the simplest explanation of why we do not find any philosophy in Egypt is that there was none.

Ancient thinkers engage in reverse plagiarism; they misrepresent their own work as coming from a more authoritative source. They live in cultures of memory where each change is condemned as a copy-error and imagination is regarded as unreliable confabulation (akin to lying). The best ideas come from a golden age. Older is better. And no civilization, in their opinion, was older than the Egyptian civilization. The Pyramid of Djoser, built 2630–2610 BC during the Third Dynasty, was a thousand years old when mammoths went extinct (about 1650 BC on Wrangel Island). Cleopatra is nearer to us in time than she was to the first pharaohs.

Admittedly, the Egyptians were secretive. For instance, their magic spells were originally restricted to the pharaohs. But after tomb robbers read the spells, they wrote them on their own coffins. Since coffin surface is limited, the poorer people had to write stolen spells on papyri. This bootleg corpus accumulated into *The Book of the Dead*.

Perhaps their priests developed philosophy and viewed it as special knowledge that could not be written down. But then why did they permit the dissemination of wisdom literature? Even the Hebrews had access. A whole paragraph of the Old Testament, Proverbs 22:17–24:22, is poached from the Instruction of Amenemope. How did the conspirators manage to keep their philosophy secret while failing to keep secret heresies, murder plots, and tomb locations?

In the case of Egyptian philosophy, absence of evidence is evidence of absence.

II

RELATIVE NOTHING

3

Lao-tzu

Absence of Action

After seeing figure 3.1 as an Indian chief, you undergo a gestalt switch to a scene in which an Eskimo peers into the black void. Something similar happened in the fifth century BC. Philosophers in China, India, Persia, and Greece turned from thinking about what exists to what does not exist.

Lao-tzu is the most complete exemplar of this shift. For, probably, he never existed.

Empty Boats

I follow those who regard 'Lao-tzu' as a pen name for sages working behind the scenes.[1] These ghostwriters abstained from becoming personalities in their own right: Taoism was developed by nobodies for nobodies. Chuang Tzu (ca. 360 BC–ca. 275 BC) writes:

> If a man is crossing a river and an empty boat collides with his own skiff, even though he be a bad-tempered man he will not become very angry. But if he sees a man in the boat, he will shout at him to steer clear. If the shout is not heard, he will shout again, and yet again, and begin cursing. And all because there is somebody in the boat. Yet

[1] Nicolas Bourbaki was invented by a group of anonymous French mathematicians in 1935 as a counterweight to their charismatic compatriot Henri Poincaré. They disapproved of Poincaré's emphasis on free-flowing intuition and analogies. Bourbaki became the most famous and influential spokesman for rigor and abstract presentation. When Ralph Boas denied that anyone was Bourbaki, a rumor was spread that Boas did not exist. I cautiously confine my comparison between 'Bourbaki' and 'Lao-tzu' to this footnote.

Figure 3.1 Eskimo face from the Illusions Index

> if the boat were empty, he would not be shouting, and not angry. If you can empty your own boat crossing the river of the world, no one will oppose you, no one will seek to harm you. . . . Who can free himself from achievement, and from fame, descend and be lost amid the masses of men? He will flow like Tao, unseen, he will go about like Life itself with no name and no home. Simple is he, without distinction. To all appearances he is a fool. His steps leave no trace. He has no power. He achieves nothing, has no reputation. Since he judges no one, no one judges him. Such is the perfect man: His boat is empty. (Merton 1969, 115)

On any practical matter, the Taoist presumption is that to do nothing is best. Specifically, conserve your energy by working with natural forces.

A fisherman at sea could return to shore by strenuously rowing. Or he might drift and hope the tide will eventually wash him up on shore.

His best way back is something in between extreme activity and extreme inactivity; catch a wave and surf back to shore.

Following the course of nature requires some conception of *nature*. The Taoist picture of reality fits the game of Go (fig. 3.2). The two players begin with a vacant board. Unlike chess, emptiness is a legal position. In chess, the empty board is only used as an abstraction, for instance, to demonstrate how the pieces move. In Go, each game must begin with emptiness. Since the original state of nature is void and the Taoist harmonizes with nature, the Taoist cultivates an empty mental state. Chuang Tzu: "The still mind of the sage is the mirror of heaven and earth, the glass of all things. Vacancy, stillness, placidity, tastelessness, quietude, silence, and non-action—this is the level of heaven and earth, and the perfection of the Tao and its characteristics" (Legge 1891, 13).

Figure 3.2 A 19 × 19 Go board model from a Sui dynasty (581–618 CE) tomb. Attribution Zcm11

This tranquil, original state is disturbed when Black places one of his black stones at an intersection of the 19 × 19 grid. White responds by placing one of her white stones at a second intersection.

When inaction is infeasible, the second best option is reaction. Lao-tzu echoes the Chinese etiquette of the guest following the lead of the host. Lao-tzu cites a military commander who played the guest during war. When his "host" advanced, he retreated.

Action tends to be self-defeating—as when the farmer pulls on a plant to hasten growth. By letting your adversary act, you allow him to defeat himself.

Instead of fighting, the Taoist yields territory. The strategy is to entice your adversary into overextending himself. Action costs energy. It thereby creates new needs. Further action becomes tempting. Supply lines thin. Your adversary can no longer sustain himself.

Victory transfers temptations from loser to victor. Lao-tzu counsels a speedy return to inaction. Upon victory, cease hostility. Do not boast. Be self-effacing. Be pacific.

From best to worst, Lao-tzu ranks omission best, then reaction, and as a last resort, action. When forced to act, one acts as little as possible, as briefly as possible, and in way that co-opts the momentum of others. Instead of blocking the wayward cart, redirect it.

With knowledge of the correct path and foresight about trajectories, a farsighted nudge obviates a larger correction. The lightness of the touch keeps the intervention inconspicuous. Feeling nothing from above, the people conclude: "*We* did it!"

The aim of Go is to encompass more territory than your adversary. White responds to Black by placing a white stone. Black and White take turns. They remove surrounded stones. When no prospect of territorial gain remains, Black and White agree to stop adding stones. They measure area. What counts is the amount of space, not the number of stones.

And only the relative amount counts. When matched against an inferior player, Go masters minimize their margin of victory. Naive losers take solace in the closeness of the score.

Archeological evidence suggests that the board began as 17 × 17 grid and expanded to the current 19 × 19 grid. One can imagine further expansions of the coordinate system that could yield a representation of any scene—rather like a black-and-white mosaic. Adding a third dimension would confer even more representational power.

Some historians conjecture that the Black and White stones represent yin and yang (fig. 3.3). The black-white contrasts signify a fundamental duality that underlies a family of opposites (fig. 3.4). The systematic nature of the distinction between yin and yang is reflected in English conjunctions. When you use 'and', you lead with the easier concept. Positive words are easier to understand and remember than negative words. Consequently, yang precedes yin in conjunctions: light and dark, heaven and earth, high and low.

Figure 3.3 Yin-yang symbol

Yang	+	Light	Heaven	High	Creative	Male	Large	Hard	Sunny
Yin	-	**Dark**	**Earth**	**Low**	**Receptive**	**Female**	**Small**	**Soft**	**Shady**

Figure 3.4 Table of dualities

The circular swirl in the yin-yang symbol hints how opposition gives rise to cycles. Each force has a tipping point at which it yields to the other.

Yin and yang must exist together—as illustrated by the need to have both women and men to sustain the species. Yin is associated with femininity and conveyed with yonic (from Sanskrit *yoni*, "vagina") symbolism.

In the West, sexualized absences are bawdy. Shakespeare's plays use the naught, 0, and the nothing gesture (thumb joining forefinger), for naughty jokes. When Hamlet asks Ophelia, "Do you think I meant country matters?" she replies, "I think nothing, my lord." Hamlet continues teasing: "That's a fair thought to lie between maids' legs." Ophelia: "What is, my lord?" Hamlet: "Nothing."

Yonic symbolism in Taoist writing is no lewder than the gender terminology for connectors and fasteners in the mechanical trades. Titillation decreases with the explicitness of the comparison to interlocking genitalia. When a plumber speaks of male and female pipe fittings, the imagery is too overt to be pornographic.

In the Taoist philosophy inspired by Lao-tzu, yin and yang arise together from an initial state of emptiness and tend to return to that state. The pond is naturally tranquil. When a stone is tossed in, crests and troughs form. The waves radiate outward from the point of disturbance until the differences even out. The pond is once again calm.

Inaction

Those writing under the name of Lao-tzu praise unobtrusiveness and noninterference. "The Tao abides in non-action, Yet nothing is left undone. If kings and lords observed this, The ten thousand things would develop naturally" (Lao-tzu 1974, 43). Lao-tzu's praise for omission and anonymity contrasts with Confucius's authoritarian emphasis on righteous action and earning the respect of one's peers and descendants.

In *The Great Learning*, Confucius says we ought to promote the greater good of humanity (animals do not count). You may seem too small to make a difference. But you are part of a chain of greater and greater wholes: individual, family, clan, village, state.

> If there is righteousness in the heart, there will be beauty in the character.
>
> If there is beauty in the character, there will be harmony in the home.
>
> If there is harmony in the home, there will be order in the nation.
>
> If there is order in the nation, there will be peace in the world. (Legge 1871, 25)

We are all in perspective when we are all in perspective. To act at these grander scales, Confucius sought power. If not as a ruler, then as an adviser. If not as an adviser, then as a teacher.

Confucius stresses the importance of correct speech because it encodes a taxonomy of roles. If you know how to describe someone (son, father, grandfather), you know how to behave toward him.

You could not have originated from anyone other than your actual parents. You are metaphysically dependent on all of your direct ancestors. They ensured you are among the fortunate few to exist. You owe them everything. Dead ancestors can only be repaid by respecting their last wishes and keeping their memory alive for future descendants. You are obliged to ensure the existence of these future admirers.

Confucius provides intricate guidance about how to behave. You should conform to your role; if you are a son, act as a son, if you are a father, act as a father, if you are leader, act as a leader. Develop good manners. Perform rituals appropriate to your station and stage of life. Be a son. Progress to being a father. Then be a grandfather! And maybe even a great-grandfather!![2]

Confucius answers the moralist's fundamental question, "What is to be done?" This question presupposes that something ought to be done. The practical side of Taoism is comprised of challenges to this presupposition.

[2] Female maxims follow male maxims as footnotes. Be first a daughter, then a mother, then a grandmother.

The Taoists are as leery of moralizing as they are of war. Once an ethicist figures out what ought to be done, he seeks power. Ethics breeds politics. And politics breeds corruption.

The corruption begins with bossiness. Even if people are genuinely misbehaving, the leader may do more harm than good by correcting them. Accordingly, Lao-tzu preaches forbearance.

Trivially, we ought to tolerate what is not wrong. Lao-tzu is bolder, advocating tolerance of some types of wrongdoing.

Can one sincerely condemn evil without being willing to do something against it? The Taoist answers that moral disapproval commits one to thwarting the misbehavior only when the benefits of interference exceed the costs. Since inflammatory acts tend to blow themselves out, the baseline for comparison is always inaction.

A modern echo of this insight is null-hypothesis testing. When evaluating the efficacy of a headache remedy, the statistician routinely compares the hypothesis that the remedy works with the hypothesis that the drug makes no difference.

The Taoist also emphasizes the possibility that someone else will intervene. One ought not to subdue a ruffian who will be more competently restrained by a neighbor. *Who* does the good deed is irrelevant.

Moralists prefer problems to be solved through high-minded motives. Taoists are less fussy. If competition rather than cooperation yields cheaper eggs, replace altruism with self-interest. In 1985, the chairman of the Communist Party Deng Xiaoping, echoed Taoist indifference when introducing free market reforms: "It does not matter whether the cat is black or white as long as it catches mice."

Taoists treasure examples of spontaneous order. But they are not so naturalistic as to endorse all natural regularities. When grasshoppers have room, they ignore each other. They abide in peaceful anarchy. When compressed in a round pen, however, the grasshoppers begin to march in a circle. Although this may seem a change of individuals into collectivists, the grasshoppers are actually being driven by the prospect of cannibalism. They move forward to avoid being consumed by the grasshopper behind them—and in the hope of gobbling the grasshopper ahead.

Plagues of locusts are orderly but not because the order serves anyone's interest. As society scales up, deleterious forms of order generate panics, mobs, and witch hunts. The proper solution is often less order. Marching damages bridges. The Taoist solution is to break stride and space out.

Taoist Resignation to Impotence

Confucius presumes all problems are solvable. Taoists counsel resignation to our limited power.

Recognition of impotence need not be frustrating. For when facts cannot be changed to fit our desires, there remains the option of fitting our desires to the facts.

Mental action is preferable to physical action because it takes less energy. If one must experiment, try a thought experiment. If one must write, then let the listener read between the lines. Put your philosophy in poetry rather than prose.

Taoist artists exploit the mind's tendency to fill in details (fig. 3.5).

> Figures, even though painted without eyes, must seem to look; without ears, must seem to listen. . . . There are things which ten hundred brushstrokes cannot depict but which can be captured by a few simple strokes if they are right. That is truly giving expression to the invisible. (Szi and Hiscox 2015, 174–75)

In Oscar Mandel's (1964) book *Chi Po and the Sorcerer* the sorcerer Bu Fu is giving a painting lesson to Chi Po. "No, no!" says Bu Fu. "You have merely painted what is. Anybody can paint what is! The real secret is to paint what isn't! Chi Po, puzzled, replies: "But what is there that isn't?"

Victorian art has been described as kenophobic (after the phobia of empty spaces). Similarly, when children paint, they fill in the entire space. Taoist art is keno*philic*.

The Politics of Nothing

Confucius attributes positive effects to virtuous causes. Rulers are the principal causes. Virtue is most important for superiors.

Figure 3.5 A page from *Jieziyuan Huazhuan* ("Manual of the Mustard Seed Garden"), a seventeenth-century Chinese painting manual

The Taoists are anarchists. The myth that there must be a top authority is due to our bias in favor of intentional explanations. We neglect rival explanations that do not involve agents.

If an authority happens to preside over good effects, we are all too eager to give credit to this salient agent. If the effects are bad, we are too eager to blame those "in control." Central authorities are more hapless than they appear.

Libertarians champion individual liberty as the highest political end. Their conception of liberty is negative: the absence of coercive interference from other agents when one attempts to do things. They extol freedom *from* censorship, taxation, and prohibitions against victimless crime. This contrasts with positive liberty: the freedom *to* eat, learn, and travel. Positive rights often conflict because of scarce resources. Scarce space makes us live close together. Now your right to sing conflicts with my right to quiet. Negative rights escape this conflict because they can be respected by doing nothing.

Most libertarians acknowledge the legitimacy of protective associations that furnish security and enforce property rights for individuals. These small associations also seek safety in numbers and so consolidate into larger associations. Eventually, the consolidation yields a minimal state. Libertarians concede these have political authority, contrary to the anarchists. Lao-tzu's praise of light, self-effacing governing styles nominates him as the first libertarian.

However, Lao-tzu does not regard the individual as strongly distinct from his environment. For the libertarian, the self marks a boundary against the demands of others. Since Lao-tzu is more holistic, he focuses less on leaders and more on ambient circumstances. Some top-down evil is conspiratorial. Some top-down stems from corruption. And much top-down evil cascades down from well-intentioned failure to foresee emergent properties.

We tend to believe conspiracy theories for bad events and to invent imaginary benefactors for good events. Even when there is nothing going on in front of us, we infer all sorts of doings behind the scenes.

Religious Taoism

Early Taoists replace somebodies with nobodies. Nobodies are in turn transformed into nothings. In the fifth century BC, religious Taoists transform nothings into nonbeings that govern nature, blurring the line between the natural and the supernatural.

On a small scale, absences inspire magical thinking. After all, absences seem to produce effects without causes. Water in a thin tube remains as long you hold your finger on top. Siphons raise water without effort. Sheets of glass adhere without fasteners. The shadow of a weightlifter mimics his effort precisely—but effortlessly. On a large scale, absences inspire awe. Canyons, caves, and the sky are sublime.

Secular Taoists resist this transformation of nothings into somethings. Granted, a riptide feels like a force that is sucking you out to sea. The swimmer has the sensation of being swallowed by a monster. But there is just an absence of pressure caused by the collapse of a sand bar. The water will flow until the pressure is equalized. A wise swimmer lets himself be carried out and then swims back after equilibrium is restored.

This inaction takes courage. It is tempting to attribute a malicious intent to the sea. Panic has an animistic metaphysics. Impersonal reversals become deliberate assaults. When a falling branch startles a dog, the animal presumes an attack. This is the policy most conducive to the dog's survival. People have the same bias toward attributing hidden agency.

Our tendency to make something out of nothing can also be abetted by our tendency to project emotions. When you peer down a cliff, the abyss appears to be drawing you down. But you are merely projecting your vertigo onto the empty space.

This does not mean that absences are harmless. Recall the dove in a vacuum chamber. The bird encounters no resistance but cannot fly. The dove soon dies. But not because of what the vacuum does. The dove perishes because of what the vacuum fails to do (provide air, counterpressure, warmth).

Absences are powerful. The error of religious Taoism is to misconstrue the power of omission as a power of commission. The pious peer over a cliff and misinterpret their vertigo as the abyss drawing them down.

Religion inflates what the secularist deflates. Lao-tzu's frugal goal of conserving energy is puffed up to the goal of increasing energy. His emphasis on having a long life balloons into a quest for immortality. The search for this ultimate elixir becomes a regulative ideal in the history of Chinese alchemy and medicine.

Absences as Causes

Fatalists believe that we have no power. We are sucked into a future state, destiny, regardless of our pasts (in contrast to determinism, which makes the past control the future). Lao-tzu is not a fatalist. Omissions are causes. When the cook fails to put out a small fire, she causes the destruction of a kitchen. A bureaucrat who can do little may still be powerful by virtue of what he can refrain from doing.

The effects of omissions are not additive. Consider a cook who has enough water to extinguish exactly one of two fires. Instead of extinguishing either fire, she does nothing. The cook's omission did not cause two kitchens to burn down. She is only responsible for one kitchen burning down. And there is no fact of the matter as to which kitchen that is.

Absences can also be *effects*. When the maid pours water on the kitchen fire, she prevents the kitchen from burning down.

Absences figure into more general puzzles of causal preemption. Consider a sink that has both an overflow hole near the top and a drain hole at the bottom. After a husband and wife discover that someone left the water flowing, the husband experiences relief: "The drain hole prevented water from rising over the lip of the sink." His wife objects "The drain hole did not prevent the flood; if the drain hole had not been at the bottom of the sink, the overflow hole would have stopped the water from spilling over the lip of the sink." The husband retorts, "But not a drop of water passed through the upper overflow hole! That hole did nothing and so did not prevent anything." Husband and wife become puzzled: surely something prevented the damage!

The Taoists believe that a wide variety of absences can be causes.

> Thirty spokes share the wheel's hub;
> It is the center hole that makes it useful.
> Shape clay into a vessel;
> It is the space within that makes it useful.
> Cut doors and windows for a room;
> It is the holes which make it useful.
> Therefore profit comes from what is there;
> Usefulness from what is not there. (Lao-tzu 1974, chapter 11)

The first objection is that counting absences as causes conflicts with the principle that causation requires the transfer of energy. When a camel collides with you, a forward oomph is absorbed. Maybe two oomphs for a Bactrian camel!

A second objection is that causation is a relation between two events. The failure of something to transpire is a nonevent. So the schema A caused B has nothing to play the role of B.

Conserving Energy

In addition to denying that he is impotent, Lao-tzu avoids fatalism by his admission that we should sometimes act. After all, he advocates acting efficiently. This is presupposed by his counsel to conserve energy. Lao-tzu warns against squandering resources on projects that are infeasible. He correctly presupposes that actions have effects.

Lao-tzu prefers actions that minimize deliberation. This ideal is on display in the story of how Lao-tzu composed his magnum opus the *Tao Te Ching*. Lao-tzu was trying to leave China. However, he was detained by the Guardian of the Pass, Yin Hsi. Perhaps acting on Mozi's doctrine that wisdom should be preserved, Yin required the sage to write down his wisdom before departing. Lao-tzu dashed off the *Tao Te Ching* in a few weeks. The Guardian of the Pass was delighted by the document and let Lao-tzu mount a buffalo and ride off into the west.

After Lao-tzu disappears, Yin Hsi becomes a recluse. Later he himself becomes a sage. Yin Hsi also disappears into the west.

Today, Confucius is reappearing in the East—thanks to subsidies from the People's Republic of China. As a boy, I first heard of Confucius through ancient Charlie Chan detective movies. The Honorable Detective would say, "Confucius say, 'Sleep only escape from yesterday'" (*Shadows over Chinatown*). "Confucius say, 'Luck happy chain of foolish accidents'" (*The Chinese Ring*). As a juvenile Sinophile, I was chagrined to learn that Confucius never actually said anything quoted by Charlie Chan. All of the wisdom emanated from 1930s Hollywood scriptwriters.[3]

[3] Yesterday's antiracist is today's racist. Charlie Chan was introduced as a hero to counter the racism caused by the villain Fu Manchu. Chan is benevolent, honorable, and

My chagrin continues. Some twenty-first-century scholars wonder whether Confucius ever said anything attributed to him—or whether he existed at all. Under one skeptical theory, Confucius was a charismatic warrior who became a quotation magnet for generations of propagandists. Cynics characterize Confucius, and indeed the very name 'Confucius', as an invention of seventeenth-century Jesuits in China. The Jesuit guests needed a Chinese sage convergent with Christianity to deflect the charge that the missionaries were pacifying the Chinese as a prelude to European invasion (more in chapter 18 on Leibniz). Confucius, according to the cynics, is a Jesuit in yellowface.

The cynics say something similar is transpiring in contemporary China. The government needs an authoritarian sage, broadly acceptable to the West, who can harmonize the yin of communist authoritarianism with the yang of capitalist individualism. The trick is to make the old imperialists eat their old bullshit.

Chinese communists eagerly await the Dalai Lama's reincarnation north of his present sanctuary in India. Communist officials plan to certify a successor who will ratify China's conquest of Tibet. They await the Dalai Lama's rebirth in Tibet. Concerned about this future kidnapping, the Dalai Lama threatens not to reincarnate. The communists deny that the Dalai Lama has any control over whether he reincarnates.

"Truth, like football—receive many kicks before reaching goal" (*Charlie Chan at the Olympics*).

sagacious (and so became a favorite of Chinese audiences). But to make Chan distinctly Chinese, the character conforms to 1930 stereotypes of the Chinese as traditional, impassive, telegraphic speakers of English. Whereas yesterday's antiracists discounted race as irrelevant, today's antiracist emphasizes race. They are offended rather than heartened by the spectacle of Caucasian actors playing Asian roles. Will the first sentence of this footnote pass the test of time? Stay tuned!

4

Buddha

Absence of Wholes

> Behold, O Monks. All compounded things are subject to decay. Strive for liberation with diligence!
>
> —Last words of Buddha

Near Kabul lies the former site of the Buddhas of Bamiyan. Before the Taliban demolished them in 2001, two monumental Buddhas had stood carved into the side of a cliff in the Bamyan valley since the sixth century (fig. 4.1). The absent Buddhas are poignant because of the Buddha's embrace of nothingness.

Idol Worship

Buddhism is extinct in the Bamyan valley. (Buddha himself prophesized that his teachings would eventually be forgotten.) Since there was no danger that the statues would be worshiped, the leader of the Taliban, Mullah Mohammed Omar, decreed in July 1999 that the statues be preserved as a tourist attraction.

Religious law tightened. A special meeting of 400 clerics ruled that the Buddhas of Bamiyan were indeed idols.

The clerics had a point. Although Buddha is famous for his doctrine that the number of selves is zero, he is depicted by an enormous number of colossal statues throughout Asia. A BIG BUDDHA makes the pilgrim feel like a little zero (0) in comparison to the big zero (0).

Of course, 0 = 0. But Buddhists draft illusions into the service of truth (fig. 4.2). No matter how carefully you measure the middle circles in figure 4.2's pair of clusters, the middle circle surrounded by

Figure 4.1 The taller Buddha of Bamiyan and its absence

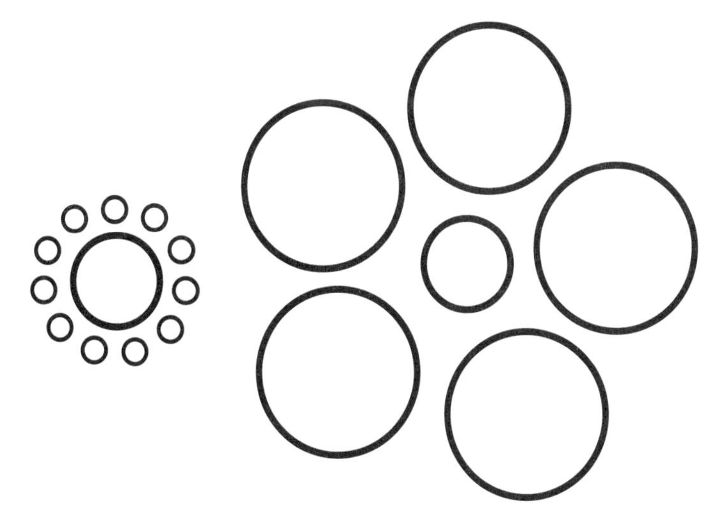

Figure 4.2 Titchener's circles

little circles looks bigger than the middle circle surrounded by the big circles.

The middle circles are of exactly the same size, but your perceptual system is isolated from your cognitive system. That is why your eyes stubbornly judge that the left middle circle is bigger than the right middle circle even after you measure. The illusion is magnified when the image pulses, slides, and rotates.

How are we to cope with the cognitive impenetrability of illusions? Buddha *co-opts* illusions. You "increase" beer in a mug by pouring the contents into a slender, cone-shaped container ∇. Your visual system attributes more volume to the tall container. The act is not fully self-deceptive because adults know, at an intellectual level, that pouring conserves volume.

As with Mohammed, Buddha denies he is divine. Buddha further denies that he is inspired by divine revelation. The gods play no role in salvation—which is merely cessation of suffering.

As with Mohammed, Buddha ordered that no images be made of him. This prohibition figures in an endearing anecdote. An artist spotted the Buddha meditating on the banks of the Ganges at Benares. The artist was so moved by the sight that he wanted to portray *something*. Yet the artist also wished to respect the Buddha's ban (as other artists did for six centuries). So the artist painted an image of the Buddha's reflection in the rippling waters of the Ganges.

The Void in Buddha's Biography

Whereas Taoists draw lighthearted support from doubts about whether Lao-tzu exists, Buddhists rebut challenges to the historicity of Buddha. Buddha existed. *Really!*

Officially, 'Buddha' is a generic title for anyone who independently learns the facts about suffering, especially if the sage then teaches these insights. But in ordinary usage, 'Buddha' is a name for the Mohammed of Buddhism, Siddhartha Gautama.

The Xerox Corporation denies their 'xerox' means 'photocopy'. They insist 'xerox' is a specific trademark name for their product. If 'xerox' is a generic term, then Xerox's rivals could describe their own merchandise as xerox machines (just as any producer of acylalicsalic tablets can now describe their pills as aspirin—formerly a trademark name). Pious Buddhists scold in the reverse direction. Instead of trying to preserve 'Buddha' as a specific name, they insist 'Buddha' is a generic noun.

Siddhartha was born in the city-state of Kapilavastu in what is now western Nepal, near the border of India. Until recently, scholars agreed he was born in 566 BCE and died 486 BCE in Kushinigar (a village so backward that Siddhartha's disciples are said to have begged him to linger long enough to die in a more dignified setting). There is now evidence that Siddhartha may have died as late as 404 BCE (Siderits 2021, 15).

Though dilute, this trickle of information has sobering implications. Kapilavastu was not a monarchy in this era. So Siddhartha could not have been a prince—as was eventually claimed centuries later. If so, all the stories erected on the premise of royalty are tall tales.

Mythmaking is a staple among Hindus. Their original Vedic religion was based on a heavenly beanstalk of tall tales. The universe is conceived as having no beginning, no end. There are cycles, each phase being dreamed into existence by gods who eventually grow bored, fall asleep, and dream up the next phase. This dreaminess reconciles diverse creation myths, thematic reversals, and contradictions.

Hinduism encourages a constructive model of worlds. Elements are combined into a panoramic mosaic. This constructive model precludes the empty world because there would be no construction. For Hindus, the classic philosophical question, "Why is there something rather than nothing?" rests on the false presupposition that absolute nothingness is possible. As with Taoism and Confucianism, there might be formlessness (Lieu 2014). But there must be something to work with.

The Chinese also leave no room for the question because of their holism. The whole is more fundamental than its parts. A world is the whole that is not itself part of any other whole. An empty world would be a composite that has no parts. But that is impossible as an ax that is missing both its blade and its handle.

Hinduism's absence of a founder is a corollary of the absence of origin. The universe has an infinite past and an infinite future. Any history with a beginning and an end shrinks into insignificance. The ensuing, oceanic casualness about the facts makes Hinduism seem more like make-believe than faith.

Initially, Hindu rituals were supposed to work by persuading the gods to intervene. But the Brahmans administering the rites eventually squeezed out the divinities. The streamlined interpretation is that the rituals work directly, more like magic spells than religious supplication to a superior agent. Failures to achieve the practical effect was blamed on the inexactitude of the wizards.

Hindu rituals accreted to mind-numbing length. Some stretch over weeks. These marathons must be performed with exquisite precision. One slip of the tongue ruins a year of preparation. What counts is the *sound* of the verses, not their meaning.

Incantations resemble computer programming except phonetics plays the role of programming syntax. The computer works by virtue of the mathematical relationships between the symbols—not their meaning. The ritual works by virtue of how the symbols sound (as with the musical incantation of Indian magician's 'Jantar Mantar Jadu Mantar'). One meaningless omission of a blank can cause the whole algorithm to miscompute. Both the priest and programmer aim at a practical goal: a good harvest, finding a spouse, curing the ill. The foreground of Vedic religion is this-worldly, using cosmology and philosophy as operatic background.

Anthropologists attribute functions to fictions. Rebirth is said to rationalize the caste structure. Everybody knows his place. A clear pecking order prevents fights. Those at the bottom are compensated by being beneficiaries of taboos that pass down property rendered impure by processes such as menstruation, birth, and death.

As befits a religion that officially denies it has an origin, there is no founding holy book such as the Koran. However, concern about the correct performance of rituals eventually led Hindu doctrines to be written in Sanskrit, the language of the gods. The point of the Vedas is acoustic fidelity (for the mantras and incantations) rather than truth. Hindus are not admitted by a catechism. Hindus trend toward polytheism, but many are monotheists. Others are agnostics or even atheists.

The priestly caste eventually assembled a corpus of poetic hymns, the Rig-Veda. Concern with metaphysical issues might be read into the Creation Hymn of the Rig Veda:

There was neither nonexistence nor existence then.
There was neither the realm of space nor the sky which is beyond.
What stirred? Where?

But the poet responsible for these lines may have been no more interested in metaphysics than contemporary songwriters. The fatalist lyrics of Jay Livingston and Ray Evans's 1956 song "Que Sera, Sera" ("Whatever will be, will be") merely sets the atmospheric backdrop for the melody. To the American audience, "Que Sera, Sera" sounds like Old World wisdom from somewhere like Spain, Portugal, France, or Italy. But the phrase has no history there and is ungrammatical in all languages. The parenthetical translation is grammatical English. But 'Whatever will be, will be' is a tautology and so says nothing about how the world is.

Around 600 BC, there was a semantic revolt against the Vedic religion. Dissidents wanted to know what it all *meant*. They did not challenge the rituals on grounds of ineffectiveness. They did question whether the effects were important.

Behaviorist religions are accused of being spiritually shallow. Reformers supplement outward practice with inward practices.

One Hindu compromise was ritualized introspection—meditation. The exercise was given a personalized goal: stripping away roles to reveal your true self.

Meditation is vulnerable to distractions that spring from the body. For instance, those sidetracked by sexual fantasies were advised to think about their mothers. A more preemptive approach was to tamp down the demands of the body through ascetic practices. Those who condemn active attack on one's body still welcome the passive dissipation of sinful desires by illness and old age.

For some seekers, meditation and asceticism were insufficiently austere. When Siddhartha became a young man, he became a renunciant, leaving his comfortable home to answer the question "Why is there suffering?"

Renunciants dropped out of the bottom of the caste system. Curiously, they still counted as a member of the moral community. Harming renunciants was thought to imperil one's village. Helping them was thought to secure spiritual advantages. This protection and support gave rise a continuum of imposters and entertainers such as snake charmers.

Renunciants sought guidance from ever more remotely situated teachers. Siddhartha is portrayed as a zealous but restless pupil who dazzles sterner and sterner gurus. His ascetic excesses come close to being fatal. At life's edge, Siddhartha has an epiphany—not the holistic "All is one" but rather the anti-holist "All is none" The whole is not greater than the sum of its parts. For parts never form a whole.

Nothingness by Means of Eliminative Reductionism

The Buddhas of Bamiyan were devilishly difficult to destroy. If the statue had been built by adding parts, it could be destroyed by disassembly. But the statue had been created by subtraction. A portion of a sandstone cliff was carved away. To destroy the statue the Taliban needed to destroy the residual part of the cliff.

Engineers had to be called in after an improvised artillery attack yielded only superficial damage. The new team proceeded top-down, recursively inserting plastic explosives into fissures formed by the higher altitude explosions. Boom, boom, boom! Bye-bye Buddhas!

The Taliban were puzzled by the international consternation. The statues were only sandstone and clay!

Siddhartha, who had forbidden statues of himself, endorses nihilism about composition. His answer to "When do parts compose a whole?" is NEVER. Siddhartha grants that common sense treats statues as being something over and above the material that composes them. According to common sense, a statue can survive loss of many parts. Common sense also endorses the reverse principle that survival of parts is not always sufficient for survival of the whole. When the Taliban completed their cycles of detonations, the parts persisted as rubble. But the statues were gone. Or so speaks the metaphysics of common sense.

Another doctrine associated with Buddhism is the impermanence of all composite things. The Taliban was only expediting the inevitable disintegration of the statues. Every complex thing falls apart.

For Siddhartha, denial of permanence needs to be couched in a conditional: *even if* composites existed, they would exist only temporarily. In fact, there are no complex things available for destruction. You cannot lose what you never had. Strictly speaking, the Buddhas of Bamiyan never existed. Consequently, it was impossible for the Taliban to destroy them.

Similarly, you do not exist and so cannot perish. Death is an illusion. Not because you will exist forever. But because you never existed.

What does exist are the simplest things. Some are physical such as the particles that are stuck Buddha-wise to compose a "statue." Other simples are mental such as the ideas that were stuck together to compose Buddha's "mind." But strictly speaking, things stuck together never compose anything new.

Buddha presupposes nominalism: everything has a position in space or time. So he would disagree with those who think that in addition to these concrete entities, there are abstract objects. For instance, mathematicians tend to assume 'two' denotes a timeless entity. The mathematicians meticulously distinguish numerals such as "2" and "II" from the number they represent. The numerals are concrete entities that have shapes and colors. What these numerals represent is a single shapeless, colorless entity that has no mass. The number two is an abstract entity that cannot repeated. There cannot be two twos.

There are infinitely many numbers. So there cannot be more concrete things than abstract things. Indeed, Georg Cantor's diagonal argument persuaded mathematicians that there are more real numbers than natural numbers. If an object is picked at random, it is infinitely more likely to be abstract than concrete.

Buddha would be surprised by the mathematician's metaphysics of abstract objects. He agrees there are ideas for counting. But these ideas are ephemera. There is the physical world, the mental world, but no third world populated by entities that have no location in space or time.

Mathematicians who came after Cantor permitted an exception to the rule that a set must have members. The null set { } has no members. All of mathematics became founded on { } by exploiting its status as necessary being that has no content.

The null set became a precedent for the more controversial *null individual*. Just as the null set is a subset of every set, the null individual is a part of every whole. This universal part is handy in formulating the logic of parts and wholes (mereology).

Buddha would reject the null set because he is a nominalist and reject the null individual because he is a *mereological nihilist*. According to Buddha, parts never combine into proper wholes. An improper part is a part that is identical to the whole (just as a set is an improper subset of itself). For the mereological nihilist, all parts are improper parts. The mereological nihilist denies the Chinese doctrine that the whole is greater than the sum of its parts.

The mereological *essentialist* maintains that an object cannot survive the loss of a part. We speak as if an elephant could survive loss of its tail. But this is loose talk adopted for the sake of convenience. The elephant's tail is as essential to the elephant as its brain.

Elements *appear* to form proper wholes when in contact or attached. Beauty also plays a role in this appearance. Rather artistically, we perceive [ ]{ }[ ] as three pairs of symmetrical brackets rather than six individual brackets.

Here is a riddle that epitomizes mereological nihilism: What is made of wood but cannot be sawed? Sawdust! For Buddha, reality is sawdust colorfully advertised as briquettes, particle boards, and myriads of other practical fabrications.

The woodcutter exists only if he is composed of hands, arms, and a head. The tree exists only if composed of roots, trunk, and branches. But all such complexes are fictions. Woodcutters and trees are akin to constellations. The Big Dipper is just a way of organizing seven stars so that they point to the North Star; two outer stars in the bowl of the Big Dipper point to Polaris. As useful as the Big Dipper may be, the constellation is merely a useful illusion. Only indivisible things exist.

When a boy builds towers from blocks, he merely arranges the blocks tower-wise. Nothing new comes into existence. So nothing passes out of existence when the tower topples. Mourning the destruction of the tower is childish because the tower was never there to be destroyed.

'Exist' derives from 'ex' (out) and 'sistere' (made to stand). When ⧈ is viewed as convex, we say a tower exists. When it is viewed as concave, we say there is only a hole. Holes do not stand out. Indeed, they recede. Mereological nihilism has a flattening effect. There is only a mosaic of atoms. None are able to stand out from the rest by teaming up into wholes. There is no absolute nothingness that can serve as a background. There is a plenum of atoms—a metaphysical democracy in which everybody is somebody and nobody is anybody.

Common sense allows for the possibility that the tower could be reassembled. This exposes common sense to the awkward question, "Does the tower persist while reconfigured as a fort?" The nihilist consistently responds by denying the presupposition that there was a tower.

Mereologists define the world as the whole composed of every atom. As a mereological nihilist, Buddha denies there is a world. There are only atoms conceived world-wise.

People have a drive for completeness—which they know how to exploit. Self-manipulative authors (I know one by introspection) circumvent procrastination by merely committing to *start* a chapter. Full commitment to finish is superfluous when the author's appetite for closure will supply missing motivation.

A drive for completeness leads us to postulate larger and larger wholes, ultimately yielding the maximal whole, the World. The same drive for completeness motivates self-construction. Instinctively, I draw a line against the environment; everything inside the line is *me*. Once begun, I want to be done. But there is no objective boundary to mark closure. Averse to arbitrariness, I try contracting. Any finite boundary is unprincipled. So I shrink to a point thumbtacked into the map of reality. But even a thumbtack has dimensions. The sculptor of the Unfinished Buddha of Borobudur plays upon this indeterminacy by leaving the right half of the statue only partly defined (fig. 4.3).

Self-Elimination

Must the nihilist retain the perceiver to have someone to succumb to the illusion? Two millennia after Siddhartha, René Descartes develops this limit on doubt in his *Meditations on First Philosophy*: previously, the skeptic had parasiticized common sense with the hypothesis that the thinker's present experiences are dreamed. Descartes parasiticized the parasite. He made skeptical hypotheses cough up certainties. According

Figure 4.3 The Unfinished Buddha of Borobudur. Attribution Okkisafire

to Descartes's method of doubt, any proposition that survived assault by all parasitical alternatives was indubitable and therefore certain.

Siddhartha's meditations proceed differently. Doubt is never portrayed as a source of illumination. Doubt is uncomfortable confusion from which one seeks relief. Siddhartha's search for his true self is simply fruitless. Like Descartes, Siddhartha invites you to look for yourself. But according to Siddhartha, all you find are particular sensations, feelings, and thoughts.

How could Descartes's "I doubt, so I am" fail to secure its conclusion? Perhaps the 'I' in 'I doubt' is a dummy pronoun as with the 'it' in 'It rained'. The semantic vacuity of this dummy 'it' explains why the word cannot be emphasized; '*It* rained' sounds odd because there is no available contrast.

When we hear 'It appears that it is noon', we do not search for a common referent for each 'it'. Nor do we search for distinct referents.

Grammar cries out for a subject. It is pacified with a dummy. According to some Buddhists, a more accurate rendering of Descartes's first premise is 'There is a doubt'. The existence of a doubter does not follow.

Another deflationary possibility is that 'I' means "the agent responsible for this very use of the word 'I'." 'I' would then purport to refer. But there would be no guarantee of success. In the Azande witchcraft culture, 'sick' resembles 'kick'. If you are kicked, then someone kicked you. Zande villagers spend their days speculating on who "sicked" who. When sick British colonialists displayed no interest in discovering their attacker, the Azande pitied the British for their illogical incuriosity.

'I' is more flexible than Descartes assumes. If my neighbor is delayed, I post a note on her door, "I will be late," to warn visitors of *her* tardiness. A criminal can postpone execution by insisting, "Traditionally, I am entitled to a last meal." When dinner is delivered, the prisoner can clarify which meal he ordered with "I am the vindaloo."

Descartes's "I doubt, therefore, I exist" is sometimes envisaged as a self-fulfilling performance. When a congressman shouts "Present!" at a roll call, he establishes his presence by the very act of declaring himself present. Saying so, makes it so.

But what can be declared into existence can be declared out of existence. In the nineteenth century, American congressmen had the privilege of declaring themselves "Absent!" to prevent a quorum. In 1890, the head of the Republican majority, Congressman Thomas Reed from Maine, outraged the Democrats by orchestrating a rule change that ended this privilege—necessitating undignified escapes and concealments.

Descartes and Siddhartha are each dualists. They agree that there are both mental things and physical things. However, Siddhartha denies that mental things depend on a mind for their existence. A defender of the self might postulate the mind as a kind of pincushion that binds thoughts. But Siddhartha adheres to a principle of lightness, preferring hypotheses that postulate fewer *types* of things—especially things that are not directly observable.

Selfless Ethics

Only mental things have intrinsic importance. Pain cries out for its own annihilation. Siddhartha's ethics coincides with negative utilitarianism: minimize suffering. Although Siddhartha agrees that happiness is good, he never endorses the classical utilitarian's principle that we should also *maximize* good. He presents the pursuit of happiness as self-defeating. Desire is a rug that is too big for a room. By shifting the rug, you appear to make progress, making one edge of the rug fit a corner, then another and another. But inevitably, other parts of the rug pucker or climb the wall. Frustration is inevitable.

If existence is worse than nonexistence, why not kill yourself? Later, the Greek pessimist Hegesias (fl. 290 BC) answered that he altruistically lingered into his eighties to teach students the advantages of suicide. He agreed with the hedonists that goal of life was happiness. But in the disheartening aftermath of Alexander the Great's conquests, Hegesias regarded the goal as impossible to achieve. He may have been influenced by Buddhist teachings that were being transmitted back from India by returning soldiers. Hegesias drives negative utilitarianism to its logical conclusion: we should strive for human extinction. The best way to prevent suffering is to prevent the existence of sufferers.

Negative utilitarianism can be made to preclude suicide by adding the assumption of an afterlife. According to Siddhartha, suicide is followed by an even worse reincarnation.[1] If you want to end your

[1] Grandmother: I believe there is no reincarnation. Grandson: Neither did I when I was your age.

existence, you need a long-term suicide plan. Specifically, you must progress through many better lives. Only after a perfect life do you get to really die.

Characterization of Siddhartha as a negative utilitarian is anachronistic. Siddhartha never uses the terminology of minimizing and maximizing. This was an innovation of the eighteenth-century British thinker Jeremy Bentham—writing just after the revolutionary application of statistics to societal questions. Bentham's disciple, James Mill, participated in the governance of India through his rising influence in the East India Company—at first marginally as a clerk, hired on the strength of his *History of India*, and then progressively with promotions culminating in his appointment as the chief administrator.

The metaphysics of the East India Company was not understood by the first leader it installed to improve the business climate in Bengal. General Mir Jafar wanted to meet the sovereign responsible for his elevation to nawab. There was no one to meet.

The East India Company's sole interest in India's past was to guide future investment. This maximizing perspective made the corporation a superlative fit for the utilitarians. The corporation has no body subject to lust or the lash. There is no taboo against the corporation choosing to end its existence when no longer profitable. The chief vice of this disembodied agent is that its moral community is restricted to shareholders. James Mill was hired to promote the profits of the few, not the many. A better boss for James Mill would have been a *cosmopolitan* corporation with every human being a stockholder. The best boss would extend the moral community to all sentient life.

James Mill discreetly applied Bentham's doctrines, trying to translate utilitarianism into the philosophical idiom of India. So did subsequent British utilitarians such as James Mill's son, John Stuart Mill (1806–1873), who injected qualitative considerations into utilitarianism and had a long career in the East India Company. Utilitarianism was in turn refined by Professor Henry Sidgwick (1838–1900). He was a member of the Council of the Senate of the Indian Civil Service Board. The engagement with India continued with Derek Parfit (1942–2017)—who explicitly elaborated

Buddhism's affinities with utilitarianism in the final chapters of *Reasons and Persons*.

British philosophers were intrigued by the inside-out manner in which Siddhartha breaks down barriers to utilitarian conclusions. The first barrier is the supreme rationality of pursuing your self-interest. This fundamental principle of economics is especially persuasive to individualists—such as the British rulers of India. For them, morality must overcome the objection that it is irrational self-sacrifice. Their fundamental question was: Why should I be moral?

Siddhartha challenges the presupposition that there is a self whose interests can be maximized. Selfishness presupposes that there is a self. So does altruism (counting others as important as *oneself*). Recognition that the self is an illusion removes a barrier to the impartial reduction of suffering. When the foot aches, the hand rubs the foot. The hand does not say, "The pain is in the *foot* and so is not my affair." All that matters is that the pain exists.

Outwardly, selfless reduction of suffering resembles the self-sacrifice of altruism. But there is no self to sacrifice. The Buddhist is as concerned with "his" suffering as much as anyone else's. Since he has the most control over his own mental states, there will be some approximation to prudence. And since he has the next most control over his kin and friends, there may be some approximation to preferring the near and dear. However, the Buddhists denial of the self removes the basis for distinguishing my pain from your pain. There is only free-floating pain.

Many admire utilitarians for their impartiality. But utilitarians are also notorious for indifference to fairness. The utilitarian does not care about *how* suffering is distributed; he only cares about how much there is. If the amount of suffering could be decreased slightly by heaping all the pain on a single individual, then the utilitarian would favor this inequality.

Each man has the greatest control over himself, and so efficiency dictates special attention to one's own case. Instead of trying to protect everyone's feet by spreading leather over the earth, the wise man puts leather on own feet.

British utilitarians believe that there are selves. Their view is that the distinction between selves is prudentially momentous—yet morally irrelevant. Selves are merely vessels into which good and evil may be poured.

These self-affirming utilitarians point out that the law of diminishing marginal utility mimics some of the effects of fair distributions. A coin relieves more suffering when in the pocket of a poor man than a rich man. But this convergence has practical exceptions. Rather than give timber to those who will only be able to build a half a bridge, we should allocate the resources to those who already possess half a bridge. The self-affirming utilitarian approves of inequality exactly when this unfairness generates the most overall well-being.

The *selfless* utilitarian neither approves nor disapproves of unfairness. He denies that fairness or unfairness is possible. Fairness is an illusion rather than a reality to be overridden.

Secret Morality and Noble Lies

Common-sense morality is offended by indifference to fairness. Utilitarianism therefore commands *secret* allegiance from its followers.

To minimize animal suffering, utilitarian activists expropriate talk of animal "rights"—despite silent agreement with Jeremy Bentham's characterization of rights as "nonsense on stilts." To combat the caste system, utilitarians tap into the moral outrage at injustice. Where moral earnestness is thought to require valuing human beings as individuals rather than vessels, the utilitarian cultivates the persona of a man who cares deeply for *you*. When the cost-benefit calculation dictates, utilitarians are obliged to become moral chameleons. Consequently, utilitarian activists send mixed signals about their real views.

The end justifies the means. What else could?

Typically, lying has worse consequences than telling the truth. False beliefs decrease efficiency at obtaining pleasure and avoiding pain. But there are many exceptions and trade-offs. Secret ballots promote sincere voting but lead to insincere reports of how one voted. James Mill argued that sincerity of votes mattered more than sincerity about reports of how one voted. His son initially agreed. Later, love intervened. Mrs. Harriet Taylor persuaded John Stuart Mill that secret ballots tipped the balance toward selfish voting rather than altruistic voting. Their commitment to truth-telling was under constant pressure until the death of Mr. Taylor in 1849.

Their commitment to truth-telling may also have been compromised by John Stuart Mill's personal and public-spirited calculation that a woman ought to be perceived as a great philosopher. *Belief* that a woman might be a great philosopher eases future recognition of great women philosophers. History progresses through the improvements of historians.

The utilitarian may also be obliged to *lie* about his moral views. The British rulers of India professed to be acting from principles that conformed to the locals they governed. Most Indians were too miseducated to comprehend utilitarian doctrines. If utilitarian principles were to be revealed, then the revelation would have to come in stages tailored to educational attainments. In addition to opportunistically accepting myths that coincide with his conclusions, utilitarian governors forge new myths.

For good or ill, this flexibility about truth-telling fits well with Siddhartha's policies. His teaching is tailored to the audience's level of understanding. For beginners, there are lists of do's and don'ts. The Three Jewels, the Four Noble Truths, the Five Precepts, the Six Realms, the Seven Factors of Enlightenment, the Eightfold Path, the Nine Consciousnesses, the Ten Perfections.

Right speech is one of the Five Precepts (a sublist of the Noble Eightfold Path). This is generally interpreted as gentle, unifying honesty. Speech should be measured by its impact on the hearers. Your words should express affection and promote concord. Right speech excludes lying, gossip, whining, and prolixity.

Like a politician, and perhaps *as* politician, Siddhartha adjusts what he says to the audience. This is often justified with the distinction between lying and omitting the truth. But the distinction does not explain the extent to which Siddhartha couches his remarks in terms of popular beliefs that he does not share.

Reservations about lying may also be tempered by an appeal to standards of the culture. Siddhartha was addressing a Hindu audience. Religious discourse was expected to be loose with the facts.

There is also a rationale that transcends culture. The duty of truth-telling varies with the burdensomeness of honesty. Language itself is biased against Siddhartha's message. He can only choose among falsehoods. Indo-European languages impose grammatical persons: first person 'I', second person 'you', third person 'he', and so forth. If Siddhartha were fastidious about not contradicting himself, he might conclude that the selfless truth is ineffable and say nothing (as he initially did after Enlightenment). Or Siddhartha might give a long-winded explanation that pedantically avoids inconsistency.

Both of these reactions oversolve the problem. Siddhartha soldiers through this superficial inconsistency, counting on his audience to charitably interpret what is said.

Siddhartha compares truth to a raft. After you reach your destination, you should not carry the raft on your head. Avoid attachment—even to the truth.

Buddha's point is to *apply* the truths. Do not study the doctrines for their own sake (like me). As Buddha sternly emphasizes, detachment from the truth is not a license to falsify the doctrines that entail that detachment. Nevertheless, the tranquil image of the sage has contributed to Buddha's becoming easily misquoted. After all, if propositions are to be asserted on the basis of their prospects for reducing suffering, then the quoter will be guided by the future rather than the past.

Now you are buttered up for joke. A Tibetan monk sees the face of Jesus in a tub of margarine. He exclaims, "I can't believe it's not Buddha!" This punchline is the title of a website entirely devoted to misquotations of Buddha.[2] The site is administered by Bodhipaksa, a member of the Triratna Buddhist Order. He treats misquotations as teachable moments. Bodhipaksa compares misquotations to genuine quotations (or at least some not so clearly fake quotations). Bodhipaksa praises some of the misquotations as wise and in agreement with Buddha. Others reflect systematic conflations with Hinduism. Bodhipaksa's tranquility is tested by Buddhists who preach toleration of misquotation. So was Buddha's (at least if Bodhipaksa has correctly quoted Buddha on misquotation).

[2] https://fakebuddhaquotes.com/i-cant-believe-its-not-buddha/.

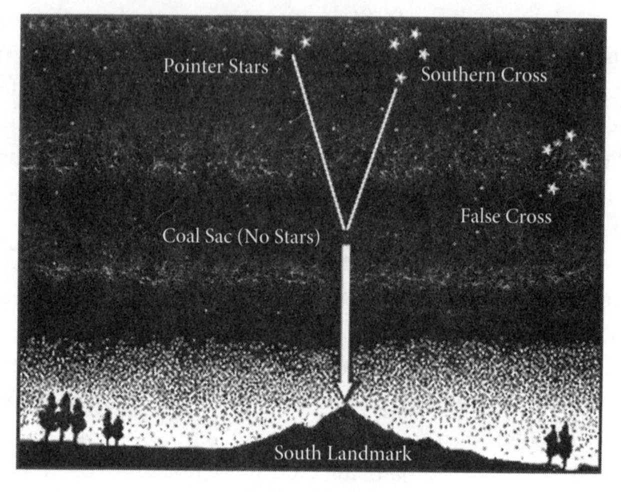

Figure 4.4 Southern Cross

Useful versus Harmful Fictions

The stars that compose a constellation are not objectively near each other. But from our perspective, the stars appear to form shapes. We "connect the dots" in various ways. Some ways are especially useful for navigation.

Consider the Southern Cross (fig. 4.4).

This is the smallest constellation that can be viewed from the Southern Hemisphere. Seeing the stars as a cross helps you find your way. You project an imaginary line along the long axis of the cross until it intersects another imaginary line formed as a T from the Pointer Stars. From that intersection, the southern celestial pole, you drop a perpendicular line to the earth. That's south.

Navigators need to avoid confusion with the False Cross and the False Pointers. These misleading figures "come into being" by resembling useful fictions.

The perceptual process spontaneously organizes parts into wholes, following principles that have proved productive in the past. This conservatism persists even when the perception is known by the perceiver to be a dangerous illusion.

The cognitive impenetrability of perceptual system is one reason why Buddhism does not aim to be purely intellectual. Buddhism is more like dieting. A fat man needs more than the insight that he overeats. He needs tips to marshal the insight.

Some of these tips exploit the very illusions that made the man fat. As people prosper, they can afford bigger plates. This makes the old portions seem smaller. Under the spell of the size-contrast illusion (recall fig. 4.2), people increase portion size. To reverse this process, dieters switch to plates that are *smaller* than their original plates. Instead of fighting the size-contrast illusion, the "self-deceptive" dieter converts the illusion into an ally.

In other-deception there is no consent from the victim. Self-deception avoids this objection.

If you cannot trick yourself into being thinner, you might settle for a thinner appearance. Black clothes make you look thinner, especially against a light background. For the visual system estimates size partly by how much light it reflects. The white Buddha looks bigger than the black Buddha (fig. 4.5).

Optical illusions even affect newly hatched chicks. In a bottom-lit cage, the hungry chicks peck off-kilter. They innately assume that the scene is illuminated from above.

Not all illusions have an innate basis. Those who learn to read English are slower to recognize WHITE when written in black ink. They have acquired an automatic ability to understand what 'white' means and so experience conflict between reporting what the word means and the color of the word.

Figure 4.5 Irradiation illusion

A convention that *was* useful can outlive its usefulness by force of habit. Perhaps castes were at one time useful for division of labor. But Siddhartha denied that they were useful any longer. His opposition to caste became a useful tool of recruitment. Lower-caste people could escape sanctions by becoming Buddhists.

Siddhartha chooses his battles. Belief in rebirth was deeply entrenched. So he does not deny the doctrine of rebirth—despite the conflict with his doctrine that there is no self. Indeed, Siddhartha co-opts the superstition by claiming to remember all of his previous rebirths! This is an infinite sequence that has no first birth (and consequently no second, no third . . .).

> Vainly I sought the builder of my house
> Through countless lives.
> I could not find him . . .
> How hard it is to tread life after life!
> But now I see you, O builder!
> And never again shall you build my house.
> I have snapped the rafters,
> Split the ridge-pole
> And beaten out desire.
> And now my mind is free. (Buddha 1976, 56)

Kutadanta, the head of the Brahmans in the village of Danamati, objected that people can be reborn only if there are people. Siddhartha compares the reborn person to a re-lit flame. While burning, the old flame persists by virtue of its continuity. When re-lit in the same lamp, the new flame is same as the old flame because it is the same type of flame used for the same purpose.

Rebirth was widely presupposed by the doctrine of karma—a bulwark of morality in India. Nowadays, we distinguish between descriptive laws of nature and normative law. A counterinstance of a descriptive law corrects the law. A counterinstance of a normative law requires correction of the infringer. Karma straddles this distinction by counting moral properties as predictive of what will happen.

Karma purports to be an *immanent*, impersonal guarantee of justice. In contrast, the British colonizers were taught to believe in a transcendent, personal guarantee. God, operating from outside the natural universe, ensures justice in an afterlife—radically unlike terrestrial existence.

For Siddhartha's contemporaries, karma answered a sequence of core questions: *Why should I be moral?* Answer: Because morality is in my own long-term self-interest. Objection: *Self-interest does not coincide with morality within a lifetime!* Reply: That just shows that there must be sufficiently many lifetimes to restore the balance. Unfair misfortunes in this life are due to my wrongdoing in past lifetimes. Failures of good deeds to advance my interest in this life are compensated by advantages that will be enjoyed in a future life.

Karma is misconceived at two levels given Siddhartha's doctrine that there is no self. First, there is no unfairness for karma to explain. Concerns about who gets what presuppose that there are selves. Second, prudence is *less* rational than morality because prudence draws a baseless boundary between my interest and the interest of others. Therefore, the karmic project of reducing morality to self-interest is a step backward, not a step forward. Karma is a nonsolution to a nonproblem.

Emotions and Make-Believe

Any fiction can be taken too far. Children who pretend to fight devolve into genuine combat. Pity that ought to focus on beggars outside a theater is directed to beggars depicted on stage.

Theater managers cultivate runaway attachments to fictions by darkening the reality around the audience to narrow attention to the stage. Participating in a fiction is analogous to running downhill. If the slope is gentle, then running gets us to our destination faster. If the hill is steep, then each brisk step forces raises commitment to a yet brisker step. We find ourselves running faster and faster just to stay upright. The initial thrill of acceleration metastases into dread.

The reminder "This is just a game" checks your descent but also deflates interest. The experienced player suspends disbelief just enough to enjoy the fruits of the fiction.

There is an emotional tension in exploiting conventions. On the one hand, we need to take the fiction seriously enough to play well. On the other hand, we need to remember that the goals of the game are objectively trivial. This gap between aspiration and reality constitutes absurdity. Perhaps the wry smile used to depict Buddha's attainment of nirvana arises from recognition of this absurdity.

Pseudoproblems

Philosophies of nothing *dissolve* problems rather than solve problems. A problem is dissolved when the interrogative sentence expressing it possesses one of the following defects: meaninglessness, ambiguity, insincerity, insignificance, false presupposition, compatibility of "rival" answers.

The spirit of the dissolutional approach is illustrated by the story of a busy man coming to Siddhartha with a problem. After the man states his problem, Siddhartha replies that he cannot solve it. The man gives Siddhartha a second chance, with another problem. Again, Siddhartha reports that he cannot solve that problem either. The man asks how Siddhartha can be the perfectly Enlightened Buddha and yet not be able solve his problems? Siddhartha tells the man that he will always have 83 problems; whenever one goes, another will take its place. Siddhartha says he cannot help with these 83 problems.

> "What can you help me with?" the man asks.
> "Oh, I can help with your 84th problem."
> "Which problem is that?"
> "That you want to solve your 83 problems."[3]

In addition to having major implications for practical reasoning, Siddhartha's denial of the self has theoretical implications. Consider how commonly philosophers formulate their problems in the first

[3] Graffiti found at the top of a wall containing the 53-meter-high Bamiyan Buddha in the 1930s: "If any fool this high *samootch* explore Know Charles Masson has been here before."

person. According to Immanuel Kant, three questions answer "all the interest of my reason": What can I know? What must I do? What may I hope? ([1781] 1965, A805 = B833). All of these questions presuppose the reality of the self. So Siddhartha thinks these questions only have corrective answers.

The Transcendental "I"

The kind of objections Siddhartha mounts to the self led Immanuel Kant to develop a compromise. Kant argues that the self is needed to explain the unity of experience. This unity underlies the spectacle of a parade. If there were no perceiver binding the perceptions together, there would only be a succession of appearances rather than an appearance of succession.

This binding agent is not equivalent to a human body. I can imagine occupying a succession of bodies. What is needed is the same type of self that makes empathy possible. I can imagine being Siddhartha fasting under the bodhi tree. But my body could not possibly have been the body of Siddhartha. So why am I not contradicting myself when I imagine myself as Siddhartha? Because this type of "I" is a frame that holds a picture.

The vertical transcendental (|) lies down to form the generality of a horizontal blank (__). Consider: "If __ is taller than Bala and Bala is taller than Chand, then __ is taller than Chand." The blank functions as a variable (For all x, x equals x) rather than an unknown (in which one is to calculate out the hidden referent 666/6 = __) or as a schema ("Put your name in the place indicated: I, ___, hereby swear to the truth"). When used as variable, __ does not designate any particular individual and is not merely used as sentence diagram. Although the open sentence '__ dreams' is neither true nor false in itself, it can be true of an individual or false of an individual. Furthermore, it can be turned into a complete sentence by adding a quantifier word such as 'For all __, __ dreams'. Since we do not colorize blanks, we use distinct letters for distinct variables.

The director and producer of *Muhammad, Messenger of Allah*, Moustapha Akkad, used the transcendental "I" to accommodate the

Muslim prohibition against pictorial representation of Islam's founder. Neither Mohammed nor his voice can be depicted on-screen. Akkad therefore filmed the scenes from Mohammed's point of view. Actors speak directly to the camera when addressing Mohammed. Tilting the camera up-down or swinging it left-right conveys Mohammed's nods. When more than a yes-no answer is required, light organ music substitutes for Mohammed's voice and the listeners repeat his words. By putting the audience in Mohammed's shoes, Akkad achieves an arresting intimacy with the Prophet.

The High Islamic Congress of the Shia in Lebanon praised the movie. In the United States, some Muslims missed the nuance about the transcendental "I." The poster for the film prominently displayed Anthony Quinn, who stars as Mohammed's uncle. The Afro-American Hinafi Muslims inferred that Quinn was starring as Mohammed himself. They seized several adjoining buildings in Washington, DC, from March 9 to 11, 1977. The protesters warned that if the film were not destroyed, they would blow up the buildings and their 149 hostages. The siege resulted in the deaths of a security guard and a radio journalist, plus the wounding of council member Marion Barry.

The siege was ended by a combination of police patience and courageous intervention by an envoy of three ambassadors from Egypt, Pakistan, and Iran. The diplomats read passages from the Koran that emphasize mercy and compassion. Eventually the gunmen surrendered.

5

Nagarjuna

Absence of Ground

Incantations must be performed the same way every time. This pressure for exact replication stimulated the creation of a stable Indic language that could be passed down through the generations. In the fifth or sixth century BC the linguist Panini formalized Sanskrit, regimenting a Brahman dialect with over four thousand rules. His perfected language became the preferred medium for scholarship and prose from 500 BC to AD 1000.

Languages indicate grammatical roles by either marking the words themselves (as in the plural *dogs*) or by marking the role with word order ('Man bites dog' is parsed differently than 'Dog bites man' despite having the same words). Latin words are so thoroughly tagged that word order is irrelevant. Whereas an Englishman keeps his pants up with a segmented belt of word order, the speaker of Latin uses syntactically marked suspenders. The writer of Sanskrit wears both a belt *and* suspenders. Sanskrit has both systems of syntactic identification.

British philosophers had to concede that Sanskrit is a safer language for reasoning than Latin or English. In English, there is a danger that the 'nothing' in 'Buddhists worship nothing' will be read as a noun (like 'Shiva') rather than a quantifier (*No* x is such that x is worshipped by Buddhists). Translating the sentence into a language with stronger syntactic signposting prevents the ambiguity from arising.

British embarrassment deepened when European languages were discovered to have originated from the same language as Indian languages. Either English regressed or at least failed to match the progress of Sanskrit. English and written Sanskrit have the same canonical word order: subject-verb-object. The kinship became apparent to the British judge and linguist William Jones. "Oriental" Jones made the eerie discovery that he could guess Sanskrit words from his knowledge of Greek and Latin!

Jones also discovered perfectly formed "Aristotelian" syllogisms in ancient Sanskrit texts. He heard them marshaled by Brahmin philosophers in oral debate. They told him Aristotle's syllogistic logic was an import from India. Aristotle's nephew Callisthenes had been the official historian for Alexander the Great's conquest of the subcontinent. Had Uncle Aristotle received more than biological specimens from India? Perhaps the Greeks forgot where they got their logic. After all, the Greeks had forgotten the location of Alexander's tomb. The location had been common knowledge because of pilgrimages by Roman Emperors. The tomb's location remains a mystery.

Study of historical India was making the ascendancy of the Britons over the Brahmins as mysterious to the Britons as it had long been to the Brahmins. Agog, Jones wondered whether European languages had regressed rather than progressed. Latin and English did appear to have become less perspicacious for the purposes of abstract reasoning. Syntactical confusions are a hazard at altitudes from which the theorist can no longer discern familiar landscape. 'Everything has a cause; therefore, there is a cause for everything' is a more tempting inference than 'Everyone has a mother; therefore, there is a mother of everyone'.

In the thirteenth century, Thomas Aquinas noted that subject-verb-object word order correlates with the action triad: agent-action-patient. Philosophers of nothing parasitize each leg of the triad. Lao-tzu draws down the energy of action. Buddha zombifies agency. Nagarjuna latches on to the patient (what we act upon). He sucks the substance out of the direct objects of action.

Buddha's skepticism about wholes gave his fellow travelers a running start. Consider an axe. The handle breaks. You replace this. The blade breaks. You replace that. Do the replacements yield the original axe or a new axe?

The question is sharpened by supposing that we reassemble the old handle and old blade. Which of the two axes is the original ax?

Siddhartha responds to these riddles by denying the presupposition that there was an axe. Rejection of composition drastically reduces his inventory of what exists. What remains are just the simplest things—atoms.

Idealism and Metaphysical Danglers

This was not simple enough. The principle of lightness led dualist disciples to trim the list to *two* kinds of atoms—mental and physical.

Two was too much for the monists. Pressing the principle of lightness even further, they asked: Is everything matter (materialism)? Or is everything mental (idealism)?

The materialist is tempted to win by throwing a shoe at the idealist. When the idealist ducks, he betrays belief in a material object.

However, the idealist can justify his dodge by appealing to the pain threatened by the trajectory of shoe. Common sense often treats touch as if it is a sure mark of the real. But there are tactile illusions. When worried that there are bugs in your bed, your skin crawls with false alarms.

Dreams differ from waking experiences in being less predictive of what will happen next. The appearance-reality distinction is a matter of bookkeeping. We wish to predict and control the flow of experience. Our attention gravitates to factors that are the best predictors. If reverie foretold the future, it would not be called "daydreaming."

The majority of Indian philosophers prefer idealism over materialism. All of our raw data is mental. So we are already committed to ideas. Should we add material things to our list of what exists? Only if they explain something that cannot be explained with ideas alone.

Astronomy illustrated how much can be explained in a purely observational way. The data consists of points of light that shift over time. The astronomer's job is to describe and predict the light patterns. He need not speculate whether the lights are gods or glowing bodies or holes in the celestial sphere. The astronomer can conceive of the stars as markers in a game such as Go.

Idealism makes ordinary things dependent on ideas. What happens to a bed when no one is present to observe it? Does the bed blink out of existence? If I am its sole observer, could I turn my head quickly and take nothingness by surprise?

The idealist answers that talk about the bed is implicitly hypothetical: *If* I have a bedroom-entry experience, then I will have a bed experience.

Wait! What *makes* this hypothetical true? The idealist cannot ground its truth in a physical room and physical bed. The hypothetical would be a free-floating truth, untethered to any piece of reality.

Absence of Grounding

Nagarjuna was a Buddhist monk who lived between AD 150 and 250 in southern India. Perhaps Nagarjuna is the sage to whom Supreme Court Judge Antonin Scalia adverts:

> An Eastern guru affirms that the earth is supported on the back of a tiger. When asked what supports the tiger, he says it stands upon an elephant; and when asked what supports the elephant he says it is a giant turtle. When asked, finally, what supports the giant turtle, he is briefly taken aback, but quickly replies "Ah, after that it is turtles all the way down." (*Raponos v. United States* 547 U.S. 715 (2006))

Nagarjuna denies any need for a foundation. He believes that metaphysical infinite regresses are benign.

These regresses struck many of his contemporaries as entailing nihilism about objects; not only would there be no axes, there would be no atoms—nothing at all. For if the existence of each object depends on the existence of another object, then there would be nothing grounding the whole chain of existence.

The monist is at one with the universe. The dualist is at two. The nihilist is at none.

Nagarjuna denies that nihilism follows from an absence of grounding. He thinks all objects could be conventional—constructions out of ever more fundamental things.

Nagarjuna's adversaries concede that constructions can be fabricated with other constructions. However, they insist that the chain of construction must eventually rest on a floor of fundamental objects.

Given that ultimate truth concerns what holds at the fundamental level, there would be no ultimate truth. Thus Nagarjuna is often interpreted as saying that there is only conventional truth.

'All truth is conventional' invites the question "Whose conventions?" Yet Nagarjuna never answers how the conventioneers get in charge of reality. This silence leads critics to stuff the following slogan into Nagarjuna's mouth: "The ultimate truth is that there is no ultimate truth."

To rescue Nagarjuna, some interpreters say he is conveying an ineffable insight. A logician will instead propose that Nagarjuna is making a *conditional* claim about what logically follows from an assumption. Or more exactly, Nagarjuna is denying a conditional that others tend to assert. Instead of arguing for anything substantial, Nagarjuna merely raises possibilities.

A philosopher is often characterized in terms of what he asserts: "Here I stand!" He is expected to write books with a title such as Bertrand Russell's *What I Believe*. But Nagarjuna regards position statements as unprofessional. What does a philosopher add to an argument when he says, "And I believe the conclusion"? Nothing from his expertise. Philosophers are trained to give reasons. Accordingly, Nagarjuna is content to exhibit what is a consequence of an assumption and what is not. His role is not to assess the endpoints of an inference—just the bridgework between premises and conclusions.

The twentieth century philosopher Gilbert Ryle is reputed to have said, "If Meinongianism isn't dead, nothing is." The logician Graham Priest (2016, xvii) begins his defense of Alexius Meinong by agreeing with Ryle. More precisely, Priest agrees with Ryle's conditionals but then infers that philosophical doctrines never die. Those who restrict philosophy to conditionals sympathicize with Priest's open-mindedness. But they would not officially make his modus tollens inference—unless 'Nothing is dead' is understood as a conditional.

At least *some* infinite regresses are harmless. The truth of 'Something exists' entails 'It is true that something exists'. Any true statement S entails a longer statement of the form 'It is true that S'. Therefore, any truth entails an infinite chain of parasitic truths. The parasites are not a burden. The trouble begins when we get revolted by these logical tapeworms and artificially dispel them. This disorients normal defense mechanisms that have accommodated to the parasites. We suffer the mental counterpart of an autoimmune disease.

Our disgust with parasitical sequences is not uniform. Zones of tolerance provide opportunities for expanding acquiescence to infinite regresses. For instance, we accept the principle that each number has a successor and *welcome* the ensuing infinite regress. If we accept this regress for the positive numbers, then, by symmetry, we should also accept the regress for the negative numbers. And if we accept negative numbers, we have a mathematical model for backward regresses. For instance, if we let the future be represented by the positive numbers and the present by zero, we can allow the past to be modeled by the negative numbers. Opening the number line liberates us from commitment to origins, such as, "Which came first, the chicken or the egg?"

Negative numbers were only rigorously discussed in Indian mathematics in the seventh century by Brahmagupta (598–668) in *Brahma-Sphuta-Siddhanta*. This is about the time Nagarjuna begins to be taken seriously. However, commercial sophistication ensured that Indian thinkers already understood Brahmagupta's analogy: "A debt cut off from nothingness becomes a credit; a credit cut off from nothingness becomes a debt." Just as there is no limit to credit, there is no limit to debt. Brahmagupta goes beyond analogy; he used negative numbers as solutions for quadratic equations and formulating the laws for multiplying and dividing negative numbers.

Brahmagupta struggles with 0. He says an integer divided by 0 is a fraction $n / 0$. But this would license the conclusion that $1 = 2$. Since any number multiplied by zero is zero, $0 \times 1 = 0 \times 2$. Division by zero yields $0 / 0 \times 1 = 0 / 0 \times 2$, which simplifies to $1 = 2$. In 830, Mahavira characterized "division by zero" as failure to divide and so concluded $n / 0 = n$. But this makes division by zero equivalent to division by one. Others assimilated "division by zero" to infinitesimal division, so $1 / n$ would be a quantity too small to be further divided.

Only with George Berkeley's critique of Isaac Newton's calculus do we get a lucid application of the distinction between meaning an absence and an absence of meaning. Berkeley exorcises infinitesimals as "ghosts of departed quantities," banning '$n / 0$' as ill-defined.

Road signs in Canada warn of extreme cold by the sign "–0." Strictly speaking, '–0' is a contradiction in terms. It is the shortest and most international of oxymorons. For negative numbers are defined as those below zero. Positive numbers are those greater than zero. So when asked whether zero is positive number or a negative number, be diplomatic and answer that zero a neutral number. The national bird of Switzerland is the Turaco. Zero should be this neutral nation's national number.

Since people prefer consistency, they will invent consistent readings of oxymorons. Some use '–0' to mark a negative number that has been rounded to zero. Computer scientists sometimes use '–0' to express negative infinity. This forges a link between 0 and ∞. Historically, negative numbers are even more paradoxical than nothing; they measure amounts less than nothing! At a party, your neighbor tells you to turn on her new audio system. The receiver registers negative 25 decibels. You wonder whether she has been tricked into buying an audio system that actually does not make any sound. Should you pretend to hear music?

After letting you squirm in a quicksand of politeness, she explains that 0 decibels resembles 0 degrees centigrade. '0 decibels' anthropocentrically marks the boundary at which human beings just begin to hear sound. By that point, birds have already been hearing ample sound.

Nagarjuna is challenging the presupposition that metaphysics concerns a fundamental level of reality. He thinks reality *might* lack a bottom level and *might* lack a top level. To show the consistency of these hypotheses, he offers models. Nagarjuna is not *asserting* that reality is topless or bottomless. He is making suppositions designed to illustrate possibilities. Nagarjuna has no *metaphysical* thesis. His thesis is meta-metaphysical: Metaphysics cannot be the study of the *fundamental* nature of reality.

How Reality Might Be Bottomless

If an object is divided and its parts subdivided, the atomist assures us that we reach a stage at which no more division is possible. For each object is ultimately composed of atoms.

Nagarjuna thinks it is possible for the division to proceed endlessly. In this scenario, each object has parts, so each thing is infinitely divisible. Contemporary metaphysicians call this type of matter "gunk." Intimations of gunk are conveyed in Jonathan Swift's "On Poetry: A Rhapsody":

So nat'ralists observe, a flea
Hath smaller fleas that on him prey,
And these have smaller fleas that bite 'em,
And so proceed *ad infinitum*.

Gunk is bad news for those who believe that parts are more basic than the whole they compose. For instance, they believe that the handle and blade of an axe are prior to the object they compose. They are happy to lengthen the chain of dependence. Iron atoms are prior to the blade. Their idea is that explanations scale up. Atomists start from the littlest things and explain the whole in terms of these parts. If there is gunk, this bottom-up explanatory strategy has no bottom and so cannot get started.

Gunk is good news for those who hold the reverse view that wholes are prior to their parts. Their explanations dangle downward from the One to the many. The Cosmos stands to its constituents as a circle stands to its semicircles. The halves are derived from a whole circle, not vice versa. Reality has a top-down direction. That's why the valves and muscles that compose the heart must be understood in terms of the whole organ. This top-down direction also explains why the organs must in turn be understood in relationship to the whole organism. And so on up to the level of species, phyla and beyond. Each thing belongs to a system of nested parts and wholes. The only thing that does not depend on anything higher is the Cosmos.

Monism comports with the religious theme that all is one. Nevertheless, the Cosmos is just another type of fundamental level. Nagarjuna would therefore say that if monism is true, it is only a contingent truth, not a necessary truth. It could be an accident that everything is material. Even if each of those material things is essentially a material thing (a bed could not have been a number or a shadow), there could have existed an immaterial thing such as a soul. So Nagarjuna would not accept a formula introduced by the logician Ruth Barcan Marcus in 1946: $(x)\Box Fx \rightarrow \Box(x)Fx$.

I recommend the Barcan formula to students seeking a philosophical tattoo. It is much less painful than the 89 character English translation: If each thing necessarily has a property, then it is necessary that each thing has that property.

How Reality Might Be Topless

The converse of gunk is knug ('gunk' spelled backward). Each thing is a part of something bigger.

To explain how marionettes move, we look up to the movements of a puppet master. If this puppet master is himself a marionette, then we look further up for the master's master. To believe in knug, is to believe there is always a bigger picture and therefore, no biggest picture.

If every combination of parts constitutes an object, then the combination of all objects would constitute a supreme-object. This supreme-object would be a counterexample to the knug principle that each thing is part of another object.

To preclude the supreme-object, the friend of knug restricts composition—not all combinations of things compose other things. To ensure that there is an infinite hierarchy of objects, he must say that there are infinitely many things. And he must also allow these things to unite in infinitely many combinations.

One way to meet all these demands is to say that any finite collection of objects composes a larger object. Wholes would be ever more comprehensive without there being any maximal whole.

This finite principle of composition would be only contingently true. For we have already seen that each object in a gunky world would be composed of infinitely many parts.

Some monists identify the world as the one thing that is not a part of anything else. They complain that the existence of knug would imply that there is no world! Nagarjuna would reply that this merely illustrates the risk of defining metaphysics as the study of the world. Metaphysics might wind up being an empty subject, as transpired for angelogy.

How Reality Might Be Both Bottomless and Topless

The logician Augustus DeMorgan supplements Jonathan Swift's gunk with knug:

> Great fleas have little fleas upon their backs to bite 'em,
> And little fleas have lesser fleas, and so *ad infinitum*.
> And the great fleas themselves, in turn, have greater fleas to go on,
> While these again have greater still, and greater still, and so on. (1915, 377)

A gunk-knug universe has both gunk and knug. Suppose each particle is composed of a miniature world—such as our own. That miniature world has particles that are themselves miniature worlds. Similarly, our world is a particle in a larger world that is itself a particle in a yet larger world.

Symmetrical and Asymmetrical Dependency

Consider a circle of lap-sitters. Each sitter depends upon another sitter. But there is no sitter upon whom all the rest depend.

Nagarjuna's doctrine of the "emptiness of emptiness" warns against inferring that the chain of dependent beings is supported by emptiness. That would betray a perverse attachment to the principle that there must be ground—making ungroundedness the ground! This tendency of reify nothing is the subject of a Buddhist joke. A merchant of an emptied shop informs a compulsive shopper that the shop has nothing. The determined shopper snaps back, "In that case, please give me some of that nothing!"

The circle of lap-sitters is kept up by mutual causation. Consider two planks that are balanced against each other: /\. Each plank depends on the other to stand. Neither has priority over the other.

A similar egalitarian dependency arises among the counting numbers. Each number gets its footing by leaning on neighbors. No number stands on its own. All stand together. None can exit to create a vacancy in the number line. Circular ungroundedness does not give us a basis for making invidious comparisons about one thing being more real than the others.

Asymmetrical chains of dependency encourage a different attitude. A ring's center of mass depends on the ring, not vice versa. This inclines us to view its center of mass as less real than the ring. The problem is not that the center of mass is invisible or unpredictable or that its predictable behavior is eccentric (such as its center of mass being located in the hole of the ring rather than in ring itself). For we have the same preference for the ring over its shadow. We see the ring's shadow but are more confident that the ring exists than its shadow.

If the ring is made of atoms, then it depends on the atoms. So we become less confident about the ring's existence. Allegiance shifts to the atoms as being the true substances.

Nagarjuna's goal is to undermine this metaphysical source of inequality. For all we know, everything might be a dependent being. There need be nothing at all that is substantial. Everything could be empty of grounding.

Even a fellow monk could feel something was missing. Nagarjuna compared the monk to a rider who is counting his herd at sunset. The rider sees one less than the morning total and so searches for a missing horse in the growing darkness. But the rider has merely overlooked the horse beneath him.

III

ABSOLUTE NOTHING

6

Parmenides

Absence of Absence

Having climbed to the precipice of being, the Chinese and Indians pan across the vista of nothingness. They see an open horizon of opportunity. The Greeks look down the Cliff. They suffer vertigo.

The Greek exemplar of understanding is geometry. To understand a sphere, focus on its intrinsic properties, such as its surface being everywhere equidistant from the center. Geometers avoid ordinary balls because their shape is distorted by the surface upon which they rest. Balls are compromised by contact with neighboring objects, wind and heat. Once isolated from environmental contingencies, however, the lone sphere confides its essential properties.

Standing on a sphere is inherently unstable. The more polished the sphere, the more difficult the balancing act. Parmenides (520–450 BC) makes his foundation a perfect sphere.

Decontextualization of geometrical objects is feasible because an object remains for examination. But when there is nothing, just the void, the examination cannot proceed. So the Greeks complain that nothingness is unintelligible.

Philosophers east of the Euphrates River think in context. They conceive of nothingness as a *relative* absence, say an absence of action (Lao-tzu) or an absence of composites (Buddha) or an absence of ground (Nagarjuna). This relativity of absences is reinforced by the practical orientation of both Chinese and Indian philosophers. To deliberate, there must be a choice between alternatives. Without this contrast, 'What shall I do?' is meaningless.

The Chinese and Indian requirement of contrast generalizes to all questions. Picture the inquirer as a waitress who presents a menu of sentences. The guest is requested to select a true sentence from the list. On this model, 'Can flamingoes bend their knees backward?' is a disguised command that means "Pick a true sentence from the following two alternatives: 'Flamingoes can bend their knees backward', 'Flamingoes cannot bend their knees backward'." The true answer, 'Flamingoes cannot bend their knees backward', prompts the waitress to ask a more complex follow-up question, 'Why do flamingoes appear to bend their knees backward?'. The waitress is requesting the guest to pick a true reason from a menu of explanations. Happily, the waitress has asked an expert who can rapidly survey potential explanations. The expert finds a correct explanation: *Because the ankles of flamingoes are located "knee high" on their legs*. The persistent waitress asks a third question, 'Why do flamingoes stand on one leg?' Exasperated, the expert answers: *Because they would fall over if they raised their other leg!*

The last answer is an insult to the waitress's intelligence. All of the potential reasons on her list exclude the flamingo's use of *two* legs. The expert relents and picks a true reason that fits her profile of acceptable reasons: 'Standing on one leg rather than two legs minimizes heat loss from submersion in water'.

Absolute absence goes unconsidered by the Chinese and Indians. They do not register 'There is nothing' as grammatical without relativization to an implicit question such as, 'What is in the bowl?'. There was no temptation to ask a question in a void.

Philosophers west of the Euphrates were tempted. The Greek philosophers are aristocrats who disdain practicality. When inheritors of the Greek legacy ask the Eternal Questions, they do not picture themselves as servants arranging alternatives for their master to select.

This does not mean that the Greeks liked the void. In fact, they appear to suffer from trypophobia (from the Greek, τρῦπα, *trŷpa*, meaning "hole," and φόβος, *phóbos*, meaning "fear"). Dimpled objects give the trypophobe goose bumps. Parmenides generalizes this disgust to holes of all sizes. Nonbeing is unmentionable, unthinkable, un-understandable.

Most Western philosophers inherit their founding fathers' aversion to nothingness. In *The Metaphysics of Morals*, Immanual Kant characterizes bastard babies as non-entities who steal into existence (Kant 1785, 109). They are not members of the moral community. Unwed mothers who kill them should not be tried for murder.

Those in the moral community who choose falsehood over truth turn themselves into non-persons (1785, 182). The liar is a no-thing who lacks even the usefulness of mere objects such as chamber pots. As a person, you are deserving of an I-Thou relationship. But when you lie, you are not even deserving of an I-It relationship.

There are a few philosophers who try to elicit sympathy with non-beings. One strategy is to raise the possibility that you are among those Kant targets, In *Nonexistent Objects*, Terrence Parsons (1980, 218) asks how you can be sure that you are not a fictional character. Any proof you adduce, such as `I think, therefore, I exist', could also be offered by a fictional counterpart. Others count up the theoretical advantages of modeling the self as an absence (Sorensen 2007).

Despite contempt for non-beings, Kant cannot bring himself to deny motion, time, and plurality. He is a negotiator who tries to moderate Father Parmenides's hard line against nonbeing.[1] Previous bargainers date back to Plato. Elizabeth Anscombe (1981, xi) aptly characterizes Western philosophy as a sequence of footnotes on Parmenides. Contemporary philosophers, such as Stephen Mumford (2021), are still trying to soften Parmenides!

Parmenides would dismiss all the departures from his hard line as incoherent compromises. For instance, if Parmenides could raise his head from Hades, he would criticize the book you are reading as contradictory (fig. 6.1). I say Hermes Trismegistus never existed. But then to whom does 'Hermes Trismegistus' refer?

I am committed to saying 'Hermes Trismegistus' means someone other than 'Vyāsa' (which purports to name the author of the Mahabharata). How could these names differ in meaning if neither has a bearer? How can I go on to affirm the triple identity: Vyāsa = Veda Vyāsa = Krishna Dvaipāyana?

In contrast to my confidence that these men do not exist, I concede that Lao-tzu *might* exist. But this is absentminded. How do I identify someone and then (in apparent amnesia of my successful search) go on to doubt that the discovered man exists?

[1] This footnote fits the glass slipper cobbled by Alfred North Whitehead's generalization, "The safest general characterization of the European philosophical tradition is that it consists of a series of footnotes to Plato" (1929, 39) .

Figure 6.1 Bust of Parmenides

The contradiction is flagrant when a denial of existence uses a pointing pronoun such as 'that' or 'this'. Suppose I direct your gaze to the giant Lao-tzu statue at the foot of Mount Qingyuan. I then declare, 'That statue does not exist!'. What a peculiar ritual! First, I establish the demonstratum for my demonstrative 'that' by demonstrating the referent to you. But then I go on to deny that the demonstratum was there to be pointed at!

Does the name 'Lao-tzu' in 'Lao-tzu might not exist' refers to the *idea* of Lao-tzu? This mental substitution backfires. If the bearer of 'Lao-tzu' is the *idea* of Lao-tzu, then 'Lao-tzu does not exist' comes out *false*. For the idea of Lao-tzu exists even more surely than his statue!

Parmenides would also criticize my description of the *destruction* of the Buddhas of Bamiyan. Destruction and generation each imply absences. The time before the Buddhas of Bamiyan came into being involved an absence of the Buddhas. The time after the destruction of the Buddhas of Bamiyan renews this absence. Thus Parmenides concludes that nothing comes into existence and nothing passes out of existence.

In my chapter on Nagarjuna, I assumed that everything is either comprised of atoms or made of gunk (stuff whose parts always have proper parts). This presupposes the possibility of plurality. Parmenides counters there must be exactly one thing. If there were two, the one thing would not be the other. But talk of "not being" is nonsense.

Parmenides's Poem

In the manner of the Chinese Taoists, Parmenides presents his philosophy in poetry. And in the manner of the Buddhists, Parmenides believes that the truth must be taught through a progression of half-truths. One of these half-truths turned out to be just as influential as Parmenides's full truth. So let us turn over this stepping stone and see what crawls out.

In the *Doxa* ("The Way of Seeming") Parmenides characterizes everything as a mixture of Light and Night. This is a reductive, black-and-white picture of reality. The basic idea is available to viewers of black-and-white television. Each figure is composed of pixels. The pixel is either on (light) or off (night). Varying the mixture of light and night approximates shades of gray. The Night-Light model of reality construes reality as binary.

The television screen creates an illusion of motion by a rapid sequence of static images. This allows a literal interpretation of Chuang Tzu's aphorism, "The shadow of a flying bird has never stirred." According to Mo Tzu (300 BC), *no* shadow can stir. Shadows exist only for an instant. They do not persist through time. The bird persists through time because one stage of the bird causes the next stage. Shadows lack this immanent causation. Instead of one stage triggering the next stage, the stages are effects of other objects. Specifically, the bird and the sun explain the sequence of shadow stages. Studying the shadow itself fails to get at the root cause of the apparent motion. Indeed, focusing on the shadow's internal history will make the shadow appear to be a magical entity that continues in motion despite collisions with walls.

We tend to be superficial in our explanations. We think the blue eyes of parents explain the blue eyes of children. But if the parents had their eyes removed, they would continue to produce blue-eyed children. The blue eyes of parents are shadows of the real cause of blue-eyed children.

Parmenides foreshadows his ultimate skepticism about motion through the half-truth of the Night-Light model. His model is not a dualist picture of reality in which there *exist* two types of things (such as ideas and bodies). Night is just the absence of light. The blackness may seem as positive as the light. But blackness is just a contrast effect. The unilluminated parts of the television screen look black because of contrast with light. When off, the screen is dull gray rather than **black**. The unilluminated pixels do not need to be infused with light absorbers such as black ink.

When a new technology is invented, old analog wine gets poured into a digital fresh bottle. We now speak of the computer representing the world in ones and zeros. Since the conceptual scheme is rich enough to represent all of reality, we wonder whether reality itself is anything more than ones and zeros (points in space that are either occupied or empty). Maybe all that we presently experience is a computer simulation!

Two things get lost in these modern translations of the Night-Light model. First, the early Greek model is spiritual. Second, the early model has yet to separate art and science. To revive this archaic perspective, we need to put aside television screens and computers, and put ourselves in the shoes of people who tried to understand reality in terms of natural resemblances.

Souls as Shadows

Every boy is struck by the resemblance between his body and his shadow. His shadow's mimicry makes the dark double systematically elusive—yet intimate. The boy's shadow always adheres to his body and yet always retreats when pursued. Unlike a receding rainbow, his shadow can be touched in the sensationless way a statue touches the ground it stands upon. Considered as a three-dimensional volume, the shadow is attached to the boy's body. Its existence is parasitic on his existence. If his body went out of existence, so would his shadow.

Despite this constant contact, the boy cannot embrace his shadow. Unlike a physical substance, his shadow offers no resistance to penetration.

In ancient Greek, 'eidolon' is defined as "any insubstantial form; image reflected in mirror or water; image in the mind, idea, phantom of the mind, fancy" (Liddell and Scott 1968). Homer also applies 'eidolon' to ghosts such as the dead Patroclus who visits the sleeping Achilles. The dead subsist in a shadowy, silent underworld, wandering listlessly. After Achilles joins Patroclus, Odysseus drips sacrificial blood to moisten the lips of Achilles's shadow. Achilles responds:

> Speak not smoothly of death, I beseech, O famous Odysseus!
> Better by far to remain on the earth as the thrall of another,
> . . .
> Rather than reign sole king in the realm of bodiless phantoms.
> (Homer 1911, ll. 488–491)

The dead survive but in such an attenuated state that they envy the living.

To modern people, the generality of 'eidolon' is surprising. For us, reflections and shadows are optical phenomena, while dreams and hallucinations are psychological. For the ancients, doubling is the essence that groups the phenomena together. The Greek term 'skia' covers shadow, "reflection, image, shade of one dead, phantom" (Liddell and Scott 1968).

Recognition of this doubling effect extends across diverse cultures, even some twentieth-century cultures. The Finnish ethnologist Uno Harva reports that

> a Lappish shaman, upon awakening from his trance, recounted that "under the earth is a people that walk with their feet against ours." If one believes further that the subterranean landscape, with its forests, mountains, rivers and lakes, reflects point for point the world above, then it is clear that *the other world is thus a mirror image of the earthly one*. In one of my studies I have shown that stated peculiarities of the Underworld seem originally to be based on *experiences with reflection in water*. Probably, also, the conception that the realm of the dead is "down below" and behind the water rests on this. (Harva 1933/38, 349)

Spirit worlds are inverted. Up is down, down is up. Left and right are reversed. Summer is winter. Rivers flow upstream. Theater inherits this reversal. Actors are referred to as shadows. You wish an actor luck by wishing him misfortune ("Break a leg").

The ancient Greeks were conversant with this reversal. In Book 2 of the *Republic* (2.359a–2.360d), Plato adapts Pindar's Myth of Gyges, in which a ring makes the bearer invisible when reversed. The curse tablets of Plato's era are written backward. Tomb figures in the late Greek period have reversed body parts.

Curiously, the inquisitive Greeks never ask of a shadow cast on a wall: "Why is my shadow reversed left-right but not up-down?" They notice that a reflection in a concave spoon fails to reverse left and right. But they do not discuss the original anomaly of reversal on a flat surface. Perhaps the supernatural status of shadows and reflections created a curiosity-curbing suspension of disbelief.

In all cultures, the vast majority of people are right-handed. This leads to the cultural universal of regarding right-handedness as good and left-handedness as bad. If you cannot tell left from right, you cannot tell right from wrong! Since the shadows of right-handed people look left-handed, shadows acquire a sinister reputation. People who resemble shadows by being left-handed are suspected of being spirits made flesh. Lefties get described with preternaturally pejorative terms: sinister, gauche, maladroit.

Because spirits are associated with shadows, they dominate as the season darkens. In the Celtic calendar, the dark half of the year is marked by the Eve of November 1 (Halloween) and runs through May Day. Yule logs were burned on the winter solstice to counter the maximal darkness.

Since any material thing casts a shadow, any object can have a spirit. Like shadows, spirits move with unnatural alacrity and in defiance of natural laws. Under a full moon, shadows cast from trees in a turbulent windy night can make the ground undulate. The traveler overcorrects, losing his balance.

Spirits are attracted to anything of the same shape as the original. Hence, you are cautioned against casting your reflection into a body of water. Spirits of the dead can be contained and preserved with look-alike statues. Mummification furnished the primary shape-home for the spirit. However, the Egyptians also provided backup statues. Since the statues resurrected by breathing, tomb robbers smashed noses to "kill" the statue.

In sum, Parmenides inhabited a culture where darkness is a spiritual kind of nonbeing. The dark side is a double realm that haunts positive reality. Parmenides's ultimate goal is a metaphysical exorcism.

Absence of Intermediates

According to the Night-Light model, underlying reality is a pseudoduality of presence and absence. The appearance of gradation and indeterminacy is an illusion. Gray hair has not a single gray hair; each hair is either black or white.

If we go by looks, there is no need to regard objects as anything more than mixtures of light and dark. As the light patterns change, figures steal into existence, grow, move through space, shrink, and flee the realm of being. Or, at any rate, there is a compelling illusion that there are figures having these adventures. In reality, there is only a shifting pattern of light and dark.

> But since all things have been named light and night
> and these in accordance with their powers [are given] to these things and these,
> then all is full of light and obscure night together,
> Of both being equal, since nothing belongs to neither. (*Doxa*, Fragment B9, in Curd 2002, 142)

Just as each representation on the black-and-white screen is a mixture of light and darkness, each object is a mixture of being and nonbeing.

Spinoza would later say that every determination is a negation—a rose is red because it is a color that is not orange or yellow or green or blue or indigo or violet. According to Hegel, "Neither in heaven nor on earth is there anything not containing both Being and Nothing" (1929, 97).

The Night-Light model of reality addresses two problems about change. According to "the way of seeming," objects come into existence and go out existence. Yet it is impossible for something to originate from nothing. It is equally impossible for something to disappear into nothing. Things can only come into existence from other things. These constituents may themselves have constituents. Ultimately there must be a fundamental layer of being that can neither enter nor exit existence.

Parmenides extends the Night-Light model to the mental realm. A man's thought is a ratio of light and night in his body. With sleep as in death (and aging), that ratio changes in direction of night.

> For according as the hot or the cold predominates, the understanding varies, that being better and purer which derives from the hot. . . . But that he [Parmenides] also attributes sensation to the opposite element in its own right is clear from his saying that a dead man will not perceive light and heat and sound because of the loss of fire, but that he will perceive cold and silence and the other opposites. And in general, all being has some share of thought. (Theophrastus, *De Sensu* 3–4, trans. John Mansley Robinson)

The dead continue to perceive—just less vividly. Perception of the dark, the cold, and silence are perceptions of what is, rather than an experience of what is not. Absences are at the low end of a hierarchy of being.

The Night-Light model reduces everything to light and night. Night is the absence of light. But an absence is what is not the case. What is not the case cannot exist. Thus, night rests on a contradiction. The Night-Light model an intermediate teaching, a preliminary to the full truth.

Parmenides presents the model more picturesquely in his proem (an introduction to a poem). He describes the journey of a young man from darkness to light. Attended by the daughters of the Sun, the young man is carried in a chariot to the temple of a goddess. She completes the poem. The goddess tells him that he must learn all things, even the unreliable opinions of men, because they represent an aspect of the whole truth.

Intermittent Existence

Geographers say "intermittent streams" feed the Evrotas River of southern Greece. These come in and out of existence with the seasons.

But where do they go when they go out of existence? Some of their water flows on from the Evrotas River to the Mediterranean Sea. Some evaporates into the clouds. Some is incorporated into organisms. But where do the ephemeral *streams* go during their periods of nonexistence?

To repair grandfather's tilting stonewall, you dismantle it for future rebuilding. If the wall has gone out of existence, then that will stymie any efforts to repair grandfather's wall! The best you can do is to cannibalize the remains of wall to create a new wall. To exist is to exist continuously. Therefore, there can be no reincarnation. Nor can there be resurrection.

You might grant that once something has gone out of existence, it must stay out. But Parmenides also challenges the assumption that something can go out of existence. When would it make this transition? Not while it is in existence. For it still exists. Not when it does not exist. For then it is already out of existence and so cannot be the subject of any changes. Since there is no point at which a thing can make the transition from being to nonbeing, destruction is impossible.

The same holds for the transition from nonbeing to being. If something does not exist, it cannot have any properties and so cannot undergo any change of properties.

Application to the Framework of Things

After applying his revolutionary principle to the things in space and time, Parmenides extends it to the framework of space and time itself. Time has three parts: past, present, and future. They cannot overlap. But any difference from one time to another would involve some earlier state of affairs going out of existence, and some later state coming into existence. But Parmenides has already shown that such transitions cannot take place, so time is unreal.

Change requires time because there must be a before and after. So change is also an illusion.

This includes motion as a special case involving change over space. Parmenides derives some special difficulties from his corollaries about space and time. Motion requires that the mover penetrate empty space. But emptiness is a kind of nonbeing. Motion also requires temporal differences. And those have already been obliterated.

Parmenides's objections to motion derive from his discoveries about negation. His disciples bred a second flock of arguments based on infinity, now known as Zeno's paradoxes.

Against Plurality

An enthymeme is an argument with an unstated premise or conclusion. Parmenides's argument against plurality appears to be an enthymeme that provides only the conclusion! He resembles a coy geometer who brilliantly deduces a conclusion and leaves it to readers to discover his premises. The clues are an assortment of potential premises. The theses Parmenides has already claimed to establish are so sweeping that several can be pressed into service.

For instance, one might use Parmenides's impossibility result against empty space: Assume, for the sake of a contradiction, that there were two distinct beings Helen and Clytemnestra. There would have to be something separating the pair from each other. The only thing that could separate Helen and Clytemnestra would be empty space. But Parmenides has already proven the impossibility of empty space. So the "two" beings must coalesce into one being.

Another conjectured argument against plurality exploits the fact that denials of identity imply a denial of existence. For instance, 'Siddhartha is not Nagarjuna' implies 'The *fact* that Siddhartha is Nagarjuna does not exist'. But Parmenides has already shown that denials of existence cannot possibly be true. So there cannot be more than one thing.

The same reasoning excludes differences in properties. Suppose there is a difference between Siddhartha and Nagarjuna. That means there must be some property, say tranquility, such that Siddhartha is tranquil and Nagarjuna is not tranquil. But when you say that Nagarjuna is not tranquil you are saying that the tranquility of Nagarjuna does not exist. If it is true that any property possessed by x is also possessed by y, then x must be identical to y. Since this is true for any x and any y, everything is identical to x. This single x is all there is. Given the task of inventorying everything, Parmenides finishes quickly: one and done!

7
Anaxagoras
Absence of Total Absences

> The Greeks are wrong to accept coming to be and perishing, for no thing comes to be, nor does it perish, but they are mixed together from things that are and they are separated apart. And so they would be correct to call coming to be being mixed together, and perishing being separated apart. (Anaxagoras, Fragment 17)

Prior to Parmenides, Thales of Miletus (born in the 620s BC) inaugurated Greek philosophy as reductionism. Instead of explaining the strange (earthquakes) in terms of the stranger (angry Poseidon stamping the ground), Thales explained the strange in terms of the familiar (collisions of drifting bodies). To explain is to reduce. Everything is explainable. So everything is reducible to something. For Thales, that something is water.

Anaxagoras (born ca. 500–480 BC, died 428 BC) continues Thales's program of demystification. Spurning supernatural "explanations," Anaxagoras says a lunar eclipse is nothing but the blockage of light by the earth. Moonlight is nothing but sunlight (reflected by the moon). Sunlight is nothing but firelight. Instead of being divine, the sun is just a fiery rock. The earth is just a mixture of rocks, water, and other familiar things. The curved shadow cast by the earth during the lunar eclipse shows that earth is round. This seems to contradict the earth being flat. But Anaxagoras engineered a settlement: the earth is a squat cylinder—a disk.

Anaxagoras was forced into a more ingenious compromise by Parmenides's objection that "A solar eclipse is nothing but a silhouetted moon" ties Thales in nots. Thales is saying that a solar eclipse is *not* the sun blackened by anger, *not* the sun being consumed a dragon, *not* this, *not* that.

In addition to challenging the coherence of the absent becoming present, Parmenides made Anaxagoras sensitive to the possibility that *reductive* monism is fallacious. 'Everything is reducible to something' does not entail that there is some single thing to which everything is reducible. After all, 'Everyone originates from a father' does not entail there is some father from whom everyone originates.

Anaxagoras's master compromise is that everything reduces to everything. Everything is composed of every type of thing, albeit in tiny amounts. Change is always quantitative, never qualitative. For change is a matter of increasing a share of something that was already present. A salt crystal grows by attracting salt from seawater. The crystal diminishes into fresh water by dissolution. Change obeys a qualitative law of conservation. No new *types* of things come into existence. No new types go out of existence. Species cannot go extinct!

Air and Nothingness

A cylinder is completely filled with water. How do you empty exactly half the cylinder? (No extra measuring instruments allowed!)

A geometrical solution: pour until the surface of the water first makes contact with the bottom. The water line of the tilted cylinder now marks half the cylinder's volume (fig 7.1).

Second riddle: Is the cylinder half full of water or half empty?

As a water monist, Thales is committed to a surprising answer: *The cylinder is half full of water; indeed, the cylinder is still full of water!* Parmenides, on purely logical grounds, would challenge the riddle's presupposition that there is emptiness.

The thesis that the half-empty glass is full of water seems patently false. You can *see* the emptiness above the water line!

Figure 7.1 Cylinder of water, tilted and untilted

Anaxagoras can let Parmenides do the talking on this point: seeing requires something to look at. If there is nothing there, you are not seeing. You are hallucinating.[1]

Yet Anaxagoras sides with common sense in thinking that something is seen. Although this visual object cannot be nothing, it can be *air*.

Common sense dismisses Anaxagoras's solution as a distinction without a difference: Air *is* nothing. To weigh something on a scale, you must add something to the pan. If nothing is in the scale's pan, you are not weighing.[2]

The Pythagorean philosophers presupposed that air is nothing in their creation story. They believed the world was brought into existence out of thin air. Our world was breathed out of extracosmic void—an infinite reservoir of air.

A creation story is incomplete if its answer to "Why there is something rather than nothing?" requires something else prior to creation. The problem is to explain how something came from absolutely nothing. If air really is nothing and air is the giver of life, then we can be breathed into existence. The idea is incorporated into the biblical creation story. After creating the world out of nothing, God gives the breath of life to Adam. The idea continues to resonate. When a winter traveler falls through the surface of a frozen lake, the air pockets formed below the ice are poignant. The bubbles preserve the drowned traveler's last breath. With the spring thaws, these remnants of life complete the return into nothingness.

Anaxagoras's Experiments

Nevertheless, there are striking phenomena that suggests that air is something rather than nothing. If you stab an apple with a straw, the

[1] Contemporary physicists concur with Parmenides about the impossibility of seeing nothing. The physicist's "vacuum" is full of virtual particles. Any approximation of "nothing" would lower the pressure and make the water boil.

[2] In fact, air weighs about one kilogram per cubic meter! Yet more surprisingly, damp air weighs *less* than dry air! Light water molecules crowd out the heavier molecules of nitrogen and oxygen that compose air.

tube collapses. But if you pinch the straw shut, the tube penetrates! One explanation is that *compressed* air makes the tube rigid.

Anaxagoras relates another demonstration that air is distinct from nothing. Turn over an "empty" cup into a bowl of water. The water does not enter the "empty" cup even when the cup is half submerged. Explanation: The air resisted the water. Confirmation: puncture the upended cup. Air hisses through the hole. A little breeze can be felt as the air is pushed out by rising water.

Air is powerful. Wind topples trees. Anaxagoras believed that our disk shaped earth is held aloft by wind upwelling from beneath it. Earthquakes are the occasional whooshes of air penetrating from below the earth to the surface.

Like Thales, Anaxagoras denies that any container can be emptied. There can only be replacement of one thing by another.

For Thales, pouring is exchanging different forms of water. As a glass of ice is heated, the glass becomes half empty of ice. But this just means the remaining half of the water has turned from solid water into liquid water. When the water boils, the liquid water becomes vaporous water. The presence of water in the air is demonstrated by condensation.

Anaxagoras thinks Thales exaggerates the flexibility of water. Although water can turn into many things, water cannot turn into fire or stars. There must be more than one type of thing.

Anaxagoras agrees with Thales's ban on transmutation. The only way to get something of one type is to start with a little sample of that type.

Anaxagoras infers that everything contains a little bit of everything. The air in the top half of the glass is not empty of water. There is merely a low predominance of water. The air also has some flesh, some gold, some wax, a little bit of each type of thing.

Seminal Metaphysics

A drop of semen contains all the ingredients needed to make a son. That son becomes a man who himself has semen. Therefore, the father's semen must contain his son's semen. And his grandson's. And so on, ad infinitum.

There must be enough semen within semen to sire an unlimited number of generations. Each man is composed of a littler man. Each man is men.

Anaxagoras is the father of homuncular explanation. How do you see? There is a little man behind your eye who looks through your eye. How does *that* little man see? There is a yet littler man who looks out the little man's eye. This explanation would be circular if there are only finitely many little men. That is why Anaxagoras postulates infinitely many little men.

A man is just a formation of things in which manly ingredients predominate. His semen also contains everything needed to make his opposite—a woman.

If a man's semen falls to the earth, a plant can absorb it. So semen must also contain plants—or at least the ingredients for making plants. A man's semen is a microcosm. Each drop reflects everything in the universe. The universe, in turn, is a macrocosm reflecting everything inside it.

Semen is just an especially pellucid illustration of "All is in all." Milk also illustrates the principle (though milk fails to exemplify the aspect of origin). The baby does not destroy the milk she drinks. The milk is dissociated into components that make up bones, skin, and other organs. Nothing can be destroyed.

Anaxagoras believed that each of us partakes in something immaterial, "Nous," which is an organizing intelligence. This cosmic mind causes all physical things to rotate. Lighter bits swirl out to the periphery. Heavy chunks collect at the center—thereby creating the earth. Such is the way of the whirled.

Nous's control over matter is the same as your mind's control over your body. When you raise your foot, your foot rises because of your intention. Yet action is more than internally initiated bodily movement. When you sit with one leg crossed over the other leg, your top foot bobs with beats of your heart. The foot's shadowing of the heart is mere behavior. Nous does not shadow anything.

The novelty of introducing an immaterial intelligence into Thales's physical picture of the universe led Anaxagoras to acquire the nickname "Mister Mind." Nous predisposes naturalists to the design argument for God. If you find a round plate on the beach, there is a possibility that it formed accidentally like a plate of shale. But it is more reasonable to infer that the plate was designed. The same applies to the clams found in the sand. Since the natural formations scale up indefinitely, there is grand design for everything.

In sum, something aloof from the swirling mass of material things must direct their distribution. Nous alone is pure. Everything else is impure—and impure all the way down.

Greek Gunk

Within the material realm, anything can turn into anything. Therefore, each thing must be a miniature reflection of everything else. This ancient reflection was continued in Benoit B. Mandelbroit's (1924–2010) fractal geometry. When asked what the middle initial *B* stands for, one mathematician answered: Benoit B. Mandelbroit.

As you decompose a thing into parts, you never reach a pure sample of anything. Everything is a mixture of mixtures. The hierarchy is bottomless.

In addition, Anaxagoras's hierarchy is topless. There is no largest thing. Structurally, Anaxagoras is a Greek Nagarjuna.

Anaxagoras is more ambitious and materialistic than Nagarjuna (who saw no explanatory advantage in postulating anything beyond ideas). Like Thales, Anaxagoras wants to explain everything. He also shares Thales's preoccupation with the material world—though this is tempered by his introduction of a cosmic mind.

Nagarjuna believes it is impossible tell the difference between reality having a bottom and being bottomless. Anaxagoras believes he can prove it is both bottomless and topless.

The details of this proof are not clear from the surviving fragments of Anaxagoras's writings. These fragments suggest that he does not count artifacts as parts or organisms. You are not made of bricks and spears. Yet he does not seem to limit ingredients to a tidy group of opposites. Instead, 'everything' is restricted to natural ingredients such as gold.

Can Anything Come from Everything?

My ring is gold because pure gold is the predominate ingredient. 'Everything is everything' precludes this natural explanation. There is no pure gold. Everything is adulterated with everything.

A mathematician might hope to explain how everything can be in everything by drawing an analogy with composite numbers. A composite number is a whole number that can be divided into parts that are other whole numbers. For instance, *eight* slices of pie can be divided among four diners by allocating two slices to each diner. *Seven* slices cannot be divided into parts that are whole numbers. Any whole number that is not composite is a prime number. Prime numbers are more fundamental because composite numbers can be divided into prime numbers. Primes are the atoms of numbers. Surprisingly, the prime numbers can be put into one-to-one correspondence with the composite numbers:

Composites: 4, 6, 8, 9, 10, 12, 14, 15, ...
Primes: 2, 3, 5, 7, 11, 13, 17, 19, ...

Yet composite numbers predominate in the sense that finite samples of the number line contain mostly composite numbers.

This arithmetic model of 'Everything is everything' would work if Anaxagoras were merely saying that each thing is composed of infinitely many parts. However, gunk also involves infinite nesting of parts within parts.

This nesting makes the integrity of the ingredients dubious. Each part is qualitatively alike in being composed of just the same mixture of the stuff. No "ingredient" can be isolated for counting by virtue of the nature of its composition. So we have yet to answer the question: "Predominance of *what*?"

The Pythagorean Pentagram

Anaxagoras's doctrine of "everything in everything" may have been inspired by the pentagram (fig. 7.2). According to Lucian of Samosata (AD 120–180) the Pythagoreans used this star-shaped figure as a badge of recognition:

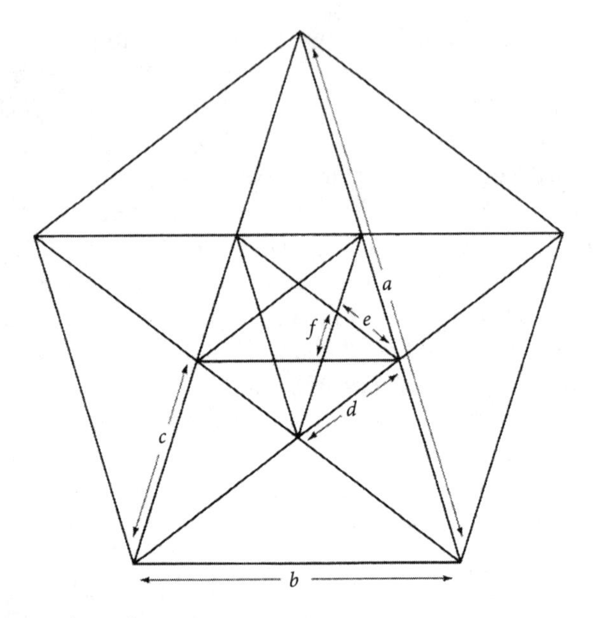

Figure 7.2 The Pythagorean pentagram

> At any rate all his [Pythagoras's] school in serious letters to each other began straightway with "Health to you," as a greeting most suitable for both body and soul, encompassing all human goods. Indeed the Pentagram, the triple intersecting triangle that they use as a symbol of their sect, they called "Health." (Lucian 1901, 179)

Pythagoras's interest in the pentagram arose from the reproductive significance his mathematical monks attached to the number five. Even numbers are female. Odd numbers are male. Since number implies plurality, the first number is two. That makes two the first female number and three the first male number. The union of this Adam and Eve is five. Five became representative of love and marriage. The Greeks represented numbers geometrically. This seems to raise an embarrassment of riches. Uh-oh—should five be represented with a five-sided figure, the pentagon, or a five-sided *pointed* figure, the pentagram? The

Pythagoreans were relieved to discover that they could have both at the same time.

Each pentagon is pregnant with a pentagram. Each pentagram is pregnant with a pentagon.

The Pythagorean pentagram provides a mathematical model of infinite, self-similar nesting. It also suggests that organisms have this pentagram-within-pentagram structure. Hypnotically, the size difference between the successive figures always equals the golden ratio. Focus on line segments in order of decreasing length—the ones marked a, b, c, d, e, f in figure 7.2. Every segment is smaller than its predecessor by a ratio of 1 to $(1 + \sqrt{2}) / 2$ (approximately 1:1.618). This is the golden ratio.

There are golden rhombuses, triangles, and angles. But the golden ratio is most clearly exemplified by the golden rectangle (fig. 7.3).

When the square component is removed, the remainder is another golden rectangle. Removing the sub-square from this miniature golden rectangle yields a yet smaller golden rectangle.

When these square removals are repeated indefinitely, the corresponding corners of the squares form an endless sequence of points. Connecting the dots yields the golden spiral (fig. 7.4).

The golden spiral inherits the self-similarity of golden rectangle; as the spiral's size increases, its shape remains the same.

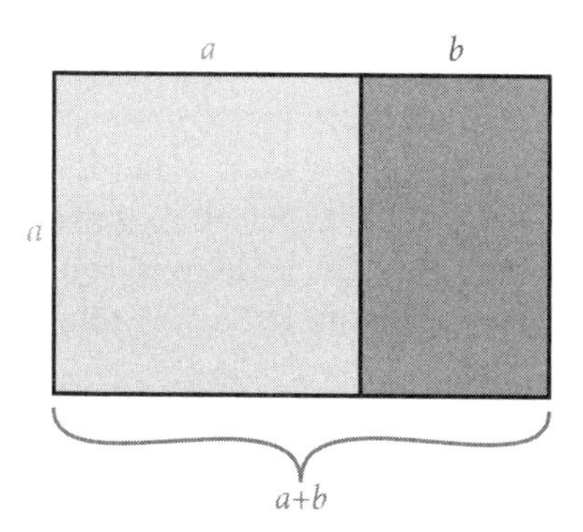

Figure 7.3 The golden rectangle

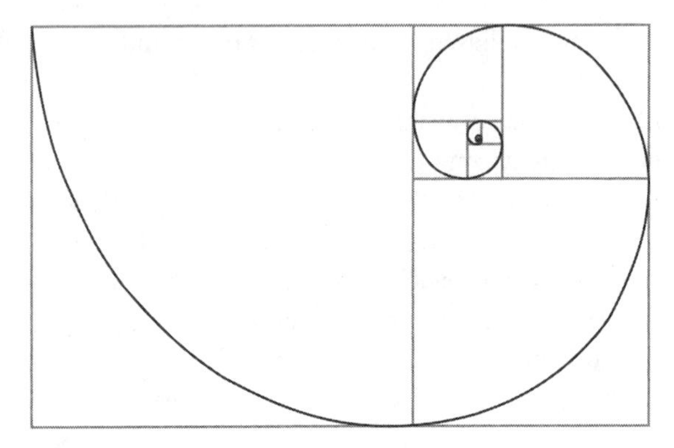

Figure 7.4 The golden spiral

The golden spiral is easy to spot in nature. At the seashore, the beachcomber finds the logarithmic pattern in the shell of the chambered nautilus. The gardener finds it in sunflowers. On mountains, the golden spiral is found in the horns of rams. Looking upward into the night sky, one can find the pattern in giant spiral galaxies. Jakob Bernoulli's epitaph is a logarithmic spiral inscribed with *Eadem mutata resurgo*; "I shall arise the same though changed."

Explaining Change

Whereas Nagarguna merely floats the possibility that everything is made of gunk, Anaxagoras is committed to the principle that everything *is* gunk. He inferred this structure as the only possible explanation of change in size and development. As Parmenides had earlier demonstrated, a thing cannot change from what it is to what it is not. Yet anything can turn into anything. So everything must contain the seeds of everything else.

A theory of change must navigate through a dilemma between destruction and survival as a mere aggregate. Succotash is a mixture of lima beans and corn. But nothing new comes into existence because the beans and corn kernels are in mere contact with each other. In order for the beans and corn to change into something, they must contribute something to the mixture and yet not get destroyed in the process.

Anaxagoras and Death

There are several references to Anaxagoras abandoning extensive property in his native town of Clazomenae in Asia Minor. According to Roman historian Valerius Maximus, Anaxagoras, coming home from a long voyage and finding his property in ruin, remarked: "If this had not perished, I would have." "This is a sentence," says Valerius "denoting the most perfect wisdom."

Was this wisdom compatible with Anaxagoras's metaphysics? If destruction is impossible, how is death possible? Just as nothing new can come into existence, nothing old can go out of it. Was his metaphysics just an instance of the psychological defense mechanism of denial?

Diogenes Laertius portrays Anaxagoras as highly conscious of death. Encountering a man who was grieving that he was dying in a foreign land, Anaxagoras consoles the man with the observation that the descent to hell is the same from every place.

Anaxagoras had been the first philosopher to settle in Athens. But the Athenians accused him of impious astronomy. Had he not denied, in writing, the divinity of the sun by reducing it to a hot rock? Anaxagoras fled to avoid the sentence no author can revise.

The death sentence later became the last word for his student Socrates. After the death of Aristotle's student and protector, Alexander the Great, Aristotle fled. Aristotle "did not wish Athens to sin twice against philosophy." Health compromised by the narrow escape, Aristotle soon died.

After learning that Athenians condemned him to death in absentia, Anaxagoras responded, "Nature has long since condemned both them and me." When he learned of the death of his two children, he said, "I knew that I had become the father of mortals."

For Anaxagoras, death is renewal. Perhaps this is the basis for my favorite death story about Anaxagoras. Esteemed in Lampsacus, the city that had given him refuge, Anaxagoras asked as a last favor that the children of the city be given an annual day's holiday in his memory.

Diogenes Laertius reports that this dying wish was granted. The holiday continued to the day of Diogenes's writing.

8
Leucippus
Local Absolute Absences

Leucippus shrinks Parmenides's One to ones (atoms) (fig. 8.1). This bottom-up metaphysics continues Thales's quest for a reductive explanation of everything. Atomism partly conforms to Parmenides's stricture against *fundamental* change. Atoms never change their *intrinsic* natures. They move but have no moving parts. Growth and destruction are confined to composites that reduce to atoms.

Leucippus is said to have originated from Miletus and from Elea and from Abdera. Diogenes Laertius cannot make up his mind (Diels and Kranz 1951, 67A1)! Despite the multiple origins, there are no mentions of Leucippus by his contemporaries. Our single direct quotation from Leucippus, "that nothing happens in vain," seems to *conflict* with the haphazardness of atomism. All the consistently atomistic doctrines attributed to Leucippus match those of his protégé Democritus (ca. 460 BC–ca. 370 BC).

Their perfect agreement raises the suspicion of reverse plagiarism. Perhaps Leucippus was invented to secure a chain of discipleship with Parmenides (who taught Zeno who taught Leucippus who taught Democritus). This suspicion is stoked by Diogenes Laertius's report that Epicurus denied the existence of Leucippus (Diels and Kranz 1951, 67A2).

Our principal source for pre-Socratic philosophy, Aristotle, assumes Leucippus exists. Some scholars say we are forced to go along with Aristotle's assumption. If Aristotle is unreliable, then we have too little documentation to support any claim to know about any philosopher who precedes Socrates. We would be like an audience who will not accept what the author says after "Once upon a time . . . ."

Internal inconsistencies and slips in Aristotle's history prevent us from legislating his story as entirely true. Aristotle is an unreliable

Figure 8.1 Parmenidean atoms

narrator. He mistakes nonexistent philosophers as actual philosophers. In Aristotle's discussion of time's passage, he voices rare approval of a Pythagorean philosopher, Paron. 'Paron' is actually a participle ("being present"), not a name. "Presence in absence" (*parōn apōn*) was a catch-phrase to convey the paradoxical way mental states render absent things "present." The phrase eventually became the title of one of John Donne's (1573–1631) metaphysical poems. First stanza:

> Absence, hear thou my protestation
> Against thy strength,
> Distance, and length;
> Do what thou canst for alteration:
> For hearts of truest mettle
> Absence doth join, and Time doth settle.

Mementos give us the footing to dangle one leg over the abyss of nonexistence. By naming your son after your father who is no more, you balance the past against the future. The shared name conveys more than its present bearer.

Larger mementos are crafted. Names compose sentences. Sentences compose paragraphs. Paragraphs compose chronicles. The retired soldier-turned-historian keeps the names of dead comrades on the lips of the living. As for foes, the old soldier may take the historian's revenge of not mentioning the rascals.

Past historians are guilty of more than sins of omission. They forge documents. Such apocrypha had to be identified and segregated from authentic records. As forgery detection methods improved, some of the documents originally intended for protection were themselves exposed as inauthentic. In the Middle Ages, rumors arose of a book, *The Three Imposters*, that alleged Moses, Jesus, and Mohammed never existed. Such was the demand for this nonexistent book that many "copies" were produced.

In 1440 Lorenzo Valla proved that the Donation of Constantine could have been written in the fourth century. The document used words unknown to Constantine such as "satrop." No one had previously thought to search the documents for anachronisms! Since this tardiness also applied to the forgers, shockingly many documents were shown to refute themselves by the very language used in their lies.

Forgers did learn not to use `Constantinople' in a document that predates the city. But they could not anticipate advances in chemistry. Each breakthrough exposed more fraud. By the nineteenth century, scholars were feeling massively duped. What they thought was history was lies. What they thought were lies was history. They were nauseated by how much history is biased, sloppy, and fraudulent. The professors commenced their dry heaves. Leucippus was expelled. "He" fit the profile of a figure conjured for propaganda.

So did Jesus. There was no independent evidence of this allegedly controversial prophet—aside from proselytizers. Perhaps Jesus had kinship with the legendary Christian king of the Indies, Prester John. In 1177 Pope Alexander III replied to a letter reporting John's desire to expel Muslims and regain the Holy Sepulcher. The correspondence was used to justify centuries of exploration to unify Christendom.

Twenty-first-century historians have recovered from the shock of deception. Instead of dismissing false documents, they are mined for nuggets of truth. Advances in document analysis have helped scholars extract ore from the slag heaps of myth and lies. They gingerly reverse the nineteenth-century verdict that Leucippus's nonexistence is more probable than his existence.

Contemporary atomists have given little notice to this improvement in Leucippus's claim to exists. Their doctrine places no weight on testimony of founders. In contrast, the rising probability of Jesus's existence is a relief to Christians. They rely on Jesus's testimony.

Atomists, such as Democritus, are committed to denying the existence of atomists: "By convention sweet is sweet, bitter is bitter, hot is hot, cold is cold, color is color; but in truth there are only atoms and the void" (Diels and Kranz 1951, 68B9, trans. Taylor 1999, 142). An atomist is nothing but a mixture of atoms. Just as the world's inventory is not increased by speaking of Leukippos as well as Leucippus, extra vocabulary for *composites* of atoms does not add to the stock of existing things.

Atomists welcome the Mindlessness of nature. They reject Anaxagoras's guiding intelligence. No Nous is good Nous.

Gorgias's Defense of Nothingness

Leucippus's metaphysics includes an ingredient more controversial than atoms: the void. Parmenides had argued that nothingness is impossible.

The Sophist Gorgias counters that nothingness is necessary. Whereas the philosophers in Thales's tradition tried to prove what was most probable, the Sophists tried to prove what was least probable. For they wished to advertise their powers of persuasion. The most improbable conclusion is that nothing exists.

Gorgias begins "On Nonexistence" with a reductio ad absurdum: Suppose nonexistence exists. Then nonexistence would both exist and nonexist. Contradiction. Second reductio: Suppose existence exists. From what did existence originate? Not from nonexistence. For Gorgias has already proved existing nonexistence is contradictory. But if existence has no origin, then it has no boundary. Existence would not be separate from what it is not, namely, nonexistence. So existence cannot exist.

Even if existence existed, continues Gorgias, existence would not be the sort of thing that can be known. For you only know your representations of reality. Any representation of something existing is compatible with it not existing.

And even if you could know that existence existed, you could not communicate this to another person. Only you can think your thoughts. So even if you knew there was something, this private knowledge could not be passed on to someone else by testimony.

To summarize, nothing exists. Even if something exists, nothing can be known about it. And even if something can be known, it cannot be communicated to others.

Leucippus's Adultery

Leucippus sprinkles Beings into Nonbeing. Aristotle reports that the atomists believe

> that the full and the empty are the elements, calling one being and the other non-being—the full and the solid being, the empty non-being (whence they say being no more is than non-being, because the solid no more is than the empty); and they make these the material causes of things. (*Metaphysics*, 1.4.985b5–10)

Leucippus agrees with the Eleatic principle that only causes are real. Since atoms and the void are equally efficacious, being and nonbeing are equally real. If you were cast into the void, you would die quickly. The void is deadly because of what it fails to do (provide heat, air, pressure), not what it does. Causation does not require transfer of energy. To shoot an arrow, the archer relaxes his grip on the drawn bowstring. He ceases to prevent the arrow from flying.

Which is more prevalent, being or nonbeing? Leucippus was not sure how densely atoms populated the universe. He was certain that the number of atoms is constant. Atoms cannot come into existence or go out of existence. They lack parts and so cannot be assembled from other things. The absence of parts also prevents them from disintegrating.

Leucippus models each atom on Parmenides's One. Atoms do not undergo internal changes. Following Parmenides, Leucippus associates the real with the permanent. Atoms are eternal. Leucippus's model of atoms therefore conflicts fundamentally with the solar system model of the atom depicted on the back of the 10-drachma coin commemorating his student Democritus (fig. 8.2). What is real is permanent. Complex objects are ephemeral and so cannot be as real as their immortal constituents.

Democritus does not require atoms to be tiny. If he had known that shooting stars are as small as a grain of sand, he might have been more optimistic about observing an atom as directly as an apple. But the atomists' explanations of how objects are seen presuppose that atoms

Figure 8.2 Front and obverse of the 1992 Greek 10-drachma coin

make other things visible without themselves being big enough to see. Given that all actual atoms are tiny, reality will be microreality.

Voids are not restricted to the miniature. Voids can be big because they do not reduce to atoms. When you see the cosmic void, you see something vast.

Leucippus pictures atoms and the void as mutually exclusive. After all, they are being and *non*being. As you add further water to the half-full cup, there is less emptiness. But the water had to come from somewhere. So emptiness must have increased somewhere else.

The amount of nonbeing atoms exclude never varies. Being and nonbeing are in a perfect, dynamic equilibrium.

Methodologically, we should only postulate as many atoms as needed to explain the phenomenon. This means we should maximize what can be explained with the void.

The atomists never consider the possibility that voids are useless side-effects of atoms failing to fill space. Instead, they expect voids to do work. Anaximander had already shown how the cosmic void could keep the earth from falling (by removing any reason for the earth to move in one direction rather than another). This big service set the stage for smaller services. Sponges show how little voids can *lift* water from damp gravel. The tiny voids in the sponge thereby *sort* liquid from solids. Nonbeing mimics being—and then takes over the same job at a lower wage. For instance, instead of insulating by adding material between interior and exterior surfaces of a wall, the carpenter may subtract material to isolate the inner wall from the outer wall. The void separating the walls of the container prevents the flow of energy from inside to outside. The emptier the space, the better the insulation.

As the atomists learned more techniques for substituting non-being for being, their universe thinned. In the first century AD, Hero of Alexandria thought of the world as mostly matter; a heap that harbors tiny voids between the grains of sand. By the fourth century, the Roman poet of atomism, Lucretius, conceived of the world as mostly void. Atoms are motes of dust in a beam of light. Later atomists would regard Lucretius as grossly *over*estimating the ratio of matter to void. They established the ratio to be more akin to that of the stars to cosmic space.

By the eighteenth century, the universe was so empty that Samuel Clarke denied that Isaac Newton's physics is materialistic. This de-atomization of atomism was expedited by the field theory of Robert Joseph Boscovich's (1711–1787). To avoid the spooky action at a distance entailed by Newton's gravity, Boscovich characterized atoms as geometrical points of force that have no size. He suggests that these forces must be aspects of a single universal force.

Dematerialization attracted religious thinkers. They delved into the analogy between God and space. Both are omnipresent, eternal, boundless, immaterial, and perfectly uniform. Positive pantheists identify God with everything that there is. Negative pantheists identify God with everything that there is not.

The history of nothing has mood swings. The future of nothing is foretold by the manic-depressive theologian: I am nobody. Nobody is perfect. Only *God* is perfect. Therefore, I am God!

No theologian consummates the divination of space. No theologian flatly asserted that God is identical to space. However, several repeated Isaac Newton's conclusion that space is an attribute of God or that space is God's sensorium.

The elimination of atoms would suit Leucippus's principle that we postulate only as many as atoms as needed to explain the phenomena. Even for atomism, the optimal number of atoms is therefore zero.

Slipping In through Cracks of Being

Atomism does not entail the void. However, geometry pressures atomists to acknowledge microvoids. For illustration, in the *Timaeus* Plato presents a version of atomism designed to preclude the void. He conjectures that the universe is completely packed with five types of atoms. Their shapes are based on the five regular polyhedra. Johannes Kepler portrays these "Platonic Solids" in his *Mysterium Cosmographicum* (fig. 8.3). Plato associates four of the Platonic solids with the four elements. The fifth Platonic solid, the dodecahedron, composes the heavens. Plato modestly calls his Plenist atomism a "likely story," not a certainty.

In "The Heavens" Aristotle protests that Plato's story is worse than uncertain; it is impossible. A mixture of all five Platonic solids cannot be snugly packed like cubes. There must be interstitial vacuums such as the gaps that would inevitably form if the atoms were balls.

Aristotle is correct. But then he goes on to falsely claim that tetrahedra can complete space. It is testament to Aristotle's authority that this claim was accepted for seventeen hundred years by mathematicians. 'Tetrahedra can complete space' is easily refuted by taking tetrahedra and trying to fit them together. These experiments always yield gaps. The tetrahedra do not even fit around a single point!

Nearly all combinations of simple shapes have interstitial voids as a side effect. As an atomist becomes more resigned to these absences, he begins to look for ways of employing them. For instance, theorists used them to explain surprising failures to conserve volume. Adding a liter

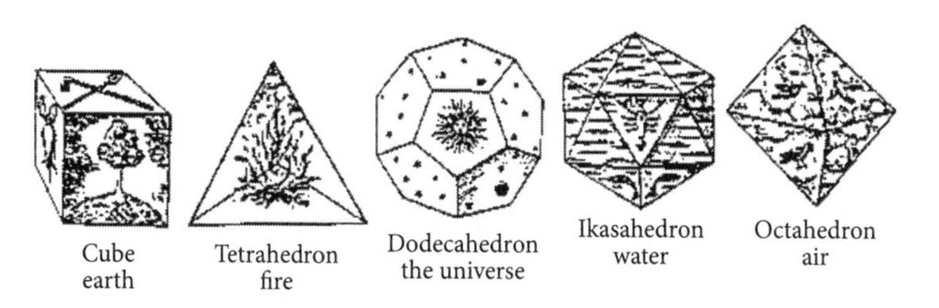

Figure 8.3 Kepler's portrayal of the Platonic solids

of alcohol to a liter of water yields less than a two-liter mixture. If one liquid has interstitial voids big enough to lodge parts of other liquid, then those parts may slip into those gaps.

Since Leucippus believed interstitial gaps were a resource rather than a liability, he did not limit atomic shapes to those that could complete space. Indeed, since there was no more reason for an atom to have one shape rather than another, Leucippus inferred atoms come in every possible shape. In principle, these atoms could fill space. But nearly all of their possible arrangements would leave gaps. The limitless invisible variety scales up to endless visible variety.

Middle-Sized Voids

Interstitial voids are not limited to the atomic realm. Interstitial gaps can be used to explain spontaneous sorting. When a basket of apples is shaken on the ride from the orchard, the smaller apples slip down between the gaps. The bigger apples rise to the top. The same principle explains why big rocks rise to the top of the soil with spring (to the perennial annoyance of farmers). Freezing and thawing shakes the earth in slow motion.

The main explanatory service of the void is to explain motion. If there were only atoms, then the atoms would block each other. Think of a universe packed tight with tiny cubes. No cube can move because there is no room to maneuver.

Anaximines addressed the gridlock by assuming that the cubes can be compressed. For Leucippus, this just raises the question of how matter can be compressed. A sponge can be compressed because it has internal spaces. It can expand because its parts can spread out so that the internal spaces become bigger.

Since the internal spaces do not weigh anything, they also explain why two objects can have the same weight and yet differ in volume. The reverse is also explained: objects with equal volume can have unequal weight. A ball of wood shavings is lighter than a solid ball of wood because the solid ball has less empty space.

Mobilizing Vacuums

Leucippus pictures the void as a kind of negative substance. If voids were the same as absolutely empty *space*, then voids would be immobile. Regions of space cannot move.

We do report the movement of holes. The hole in my ring moves as the ring rolls across the floor. This change of place implies that a hole is not a volume of space.

The same point can be made for vacuums. Picture a vacuum in a sphere being tossed from A to Z. The vacuum starts at A then travels to the intermediate position B then C then . . . then Y then Z.

A vacuum immobilist could continue insist that a vacuum is just a volume of space by redescribing the tossed sphere as containing a sequence of twenty-six distinct vacuums. The vacuum at A gets replaced by the vacuum at B, which in turn gets replaced, by the vacuum at place C, and so on down to Z. The immobilist will require more of motion than a shift in location. An earlier stage of the object must cause its next stage ("immanent causation"). Vacuums lack the integrity needed to preserve their identity over time.

The immobilist reinforces his case by spinning the sphere (Lewis and Lewis 1970). The vacuum within the sphere does not rotate. If you think it does rotate, consider a sphere suspended within a larger sphere. The outer sphere spins clockwise. The inner sphere spins counterclockwise. Do the inner and outer vacuums rotate in opposite directions? No! So a vacuum is incapable of rotational motion. But if a vacuum is capable of translational motion from A to Z, then it should also be capable of rotational motion. So vacuums are just as stationary as predicted by hypothesis that a vacuum is just an empty region of space. A vacuum is an immoveable "object" that can resist any force that tries to dislodge it.

The Cosmic Void

If the amount of matter were finite, an infinite ocean of nonbeing would surround being. This ocean would be a Universal Cosmic Void surrounding the Island of Being.

This seemed unlikely to Leucippus. The void offers no resistance to movement. So the atoms in the Island of Being would disperse into the ocean of nothingness. Indeed, since the void offers no resistance, the rate of dispersal would be of unlimited speed. Atoms would instantaneously become so dilute that they could not congregate into complex objects.

Those averse to an infinity of atoms suggested that the atoms might be held together by mutual attraction. This would require action at a distance. Mutual attraction also oversolves the problem of encouraging atoms to clump together. For now all the atoms would collect into a huge dense ball.

To avoid this collapse, one would need to postulate infinitely many atoms that are, on average, evenly distributed through the universe. The mutual attractions would then cancel out.

The necessary eternality of atoms was also needed to explain why the universe is still around. Suppose each atom has a tiny chance of popping out of existence. Given an infinite past, this small possibility would have become actual. Each atom would have disappeared long ago. Yet here we are. Consequently, each atom must be a necessary being that has *no* chance of perishing. The immortality of atoms is certain.

This empirical certainty was supplemented with a definitional argument. 'Atom' is derived from the Greek "uncuttable"; they are what is left over after all possible cutting. These minimal units cannot be destroyed in the way ordinary things are destroyed—by being taken apart.

The atomists do not address the immortality of the *void*. If the void were just space, then it could co-exist with the object that occupies it. The void would then be indestructible. However, if the void is defined as *empty* space, then any object that moves into the void interrupts its existence. Over time, some trespassing atom would invade each spot in space. Given an infinite past, the void would be erased and re-erased infinitely many times.

Opposite Paths to the Extracosmic Void

The systematic way to argue for the void is up from a bottom of tiny voids. First, show that interstitial voids are forced by almost any choice of atom shape. Since little voids are almost inevitable, middle-sized voids should be possible. And if middle-sized voids are possible, then so are large voids.

This bottom-up approach to the extracosmic void assumes atomism. Like Anaxagoras, the Stoics rejected atoms. Any void within the cosmos was unacceptable to the Stoics because the gaps threatened their view of the cosmos as a unified organism. For instance, if there were a void surrounding the earth, then the sun would lack a medium by which to transmit light and heat to the earth. All causation is by contact, either direct or indirect. A void around the earth would work like the vacuum between the double lining of a perfect thermos bottle, completely insulating the earth from the sun's warmth. The enveloping void would be a giant monster who freezes the earthlings.

The Stoics regarded the extracosmic void, in contrast, as harmless. For the extracosmic void does not insulate the cosmos from anything. Although the cosmic void serves no function for the Stoics, there is a famous thought experiment that commits them to it. If space were finite, there would be an edge to it. An extremist at the frontier could put his hand through it. That is absurd, so space must be infinite.

The Stoics did not think small voids are possible. This impossibility explains adhesion between flat plates. Once the disks come into contact, separating them would require the formation of a vacuum. The plates would be permanently stuck!

Ordinary plates are separable and so cannot be contact. Some thinkers postulated a thin film buffering the plates. But if the film is there, it is everywhere. This film between every pair of surfaces has striking implications. No man walks on the earth! Instead, all men walk on the film. Everybody is an untouchable. We are all on film!

If the Chinese had known about the practical implications of atoms, they may have taken this bottom-up path to nothingness. But the Chinese read the universe top down (fig. 8.4).

You miss the redundant OF in figure 8.4 because the sentence has primacy over its words. In turn, words have primacy over their letters. In the "Chinese" font, the *H* in WHOLE is indistinguishable from the *A* in PART. When *H* and *A* are parts of whole words, however, they are distinguishable.

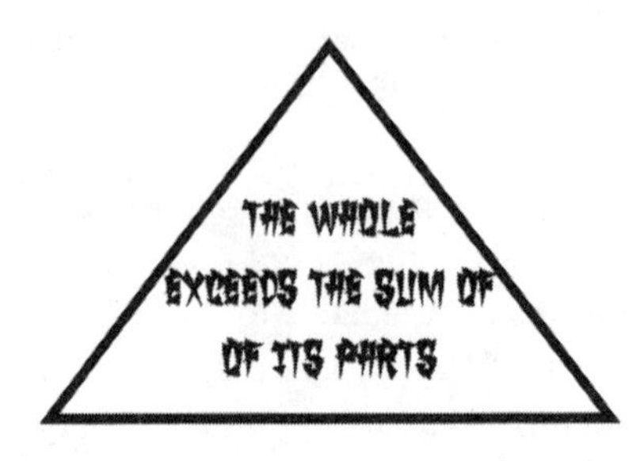

Figure 8.4 Superiority of the whole over the part

The priority of words over letters allows us to use alphabets that have fewer letters than phonemes. English tourists notice this from buildings with inscriptions on Roman monuments. The inscriptions look misspelled because Latin uses the same letter for V and U. Here is a parallel TRVTH: English inscriptions appear misspelled to Spanish speakers because English lacks the letter *ñ*.

Holism had a monopoly in China. This is why there is no indigenous Chinese atomism. In India, combinatorial thinking was common. Sanskrit has an alphabet. Letters make words. Words make sentences. A Sanskrit reader recognizes a letter more quickly in the context of a word. But his bottom-up approach will make him resist the conclusion that the word is at a more fundamental level of reality than the letter. Sanskrit makes Indians read in a combinatorial structure to reality. Linguistic atomists welcome the chemistry of the periodic table and the biology of genes.

Holism is the belief that the whole is more fundamental than the part. This outlook made atoms alien to Chinese culture.

Atomism was eventually imported to China from India. Some speculate that atomism was also imported from India to Greece. According to Diogenes Laertius, Democritus went to India to learn from the gymnosophists.

Buddha, who may have lived at the same time as Leucippus, believed there were both mental atoms and physical atoms. Buddha is far more interested in the mental atoms since his primary mission was the waning of woe. Leucippus believes there are only physical atoms and the void. His main interest is phenomena such as absorption, mixing, adhesion, and suction.

Buddha's atomism does not invoke the absolute nothingness of the void. There is only relative absence. In fact, Buddha appears to be a plenist. Atoms fill space, leaving no absolute gaps, only relative gaps.

Infinite Return

The early atomists believed that the infinite past implied an infinite return. Everything that happens has happened before. In the past, a duplicate of you read a duplicate of this book in a duplicate environment. This duplicate was preceded by an earlier duplicate. In the future, a duplicate of you will read a duplicate of this book. Your descendant duplicate will think about you and your infinite lineage of duplicates.

Your descendant will replicate your reasoning: *There are only finitely many things and finitely many properties. A state of affairs is a permutation of these things and properties. There can only be finitely many state affairs. Therefore, they must repeat. And if they repeat once, they must repeat again.*

Some atomists interpreted these repetitions as showing that complex things are also immortal. Believers in reincarnation might seize upon the combinatorial argument as a mechanism.

But most atomists distinguished between themselves and their duplicates. In any case, the atomists do not intend the immortality as any solace against the fear of death. For none is needed. Fear of death is irrational. As we shall see in the chapter on the Roman Lucretius, fear of death is to be extinguished by exposing its fallacies.

Having trouble remembering the difference between Leucippus and Lucretius? So did nineteenth-century scholars. Perhaps this phonetic rivalry contributed to the near death of 'Leucippus'.

IV
POTENTIAL NOTHING

9

Plato

Shades of Absence

The Greeks regarded the shadow as a lowly thing. Their attitude is epitomized by a tale told about the Greek orator Demosthenes (384–322 BC):

> When once at a meeting of the Athenians they would not suffer him to speak, he told them he had but a short story to tell them. Upon which all being silent, thus he began: A certain youth, said he, hired an ass in summer time, to go from hence to Megara. About noon, when the sun was very hot, and both he that hired the ass and the owner were desirous of sitting in the shade of the ass, they each thrust the other away,—the owner arguing that he let him only his ass and not the shadow, and the other replying that, since he had hired the ass, all that belonged to him was at his dispose. Having said thus, he seemed to go his way. But the Athenians willing now to hear his story out, called him back, and desired him to proceed. To whom he replied: How comes it to pass that ye are so desirous of hearing a story of the shadow of an ass, and refuse to give ear to matters of greater moment? (Plutarch 1914, 5:viii, 52)

'The shadow of an ass' was a cliché for something of trivial importance. Aristophanes (ca. 446–ca. 386 BC) uses it in the *Wasps* (191). Plato (429–347 BC) alludes to the cliché in the *Phaedrus* (260c).

Shadows were emblematic of deception. In Aesop's (620–560 BC) fable "The Dog and the Shadow," a dog is crossing a bridge with a piece of meat in mouth. He sees his shadow in the water and mistakes it for another dog with a piece of meat double the size. The self-envious dog opens his mouth to filch the larger meal and lunges for the shadow.

His real meat drops into the stream and is swept away. Aesop's moral is "Beware lest you lose the substance by grasping at the shadow."

As the fight over the ass's shadow demonstrates, shadows can be genuine objects of desire.[1] And not just by people. Fish desire shade. Tricolor herons shade the water with their wings to attract prey.

Puppeteers exploit the resemblance between objects and their shadows. Shadow theater provides the infrastructure for the allegory of the cave:

> And now, I said, let me show in a figure how far our nature is enlightened or unenlightened: —Behold! Human beings housed in an underground cave, which has a long entrance open towards the light and as wide as the interior of the cave; here they have been from their childhood, and have their legs and necks chained, so that they cannot move and can only see before them, being prevented by the chains from turning round their heads. Above and behind them a fire is blazing at a distance, and between the fire and the prisoners there is a raised way; and you will see, if you look, a low wall built along the way, like the screen which marionette players have in front of them, over which they show the puppets. (*Republic* 7.514b–520a)

Since the prisoners have been raised in the cave, they mistake the shadows for reality. The "wiser" prisoners are those who are more adept at recalling and predicting the course of the shadows. Their disputes are akin to those Demosthenes recounts between the owner and the renter of the ass.

As a playwright, Plato was familiar with the analogy between shadows and fictional characters. The characters are copies of people. Sometimes the copies have high fidelity. But in other cases the characters are caricatures—willfully distorted.

After encountering Socrates, Plato became persuaded that the physical world is itself stage. Plato wished to discover what lay behind the scenes.

[1] Ass holes are also desired (Lundgren et. Al. 2021). In arid environments, feral donkeys dig two meters to reach water beneath empty stream beds. The wells from these smart asses become contested resources among 57 species.

The shadowing relation is the cement of Plato's universe (more precisely, the two-world universe he favors prior to writing the *Timaeus*). Forms are the abstract, hidden, reality that philosophers must struggle to understand. Greek astronomers had already deduced the three-dimensional shape of the earth from the two-dimensional shadow it casts on the moon during lunar eclipses. Biophysicists revived the approach in the 1950s to deduce the three-dimensional structure of molecules from X-ray shadows cast on photographic plates. Patiently combining the snapshots allowed biophysicists to deduce that a double helix must have been casting the microshadows.

Concrete instances of a form trigger recognition of that form. If our minds began like blank slates, then inquiry would be hopeless. Even if our question were correctly answered, we would never be in a position to recognize the answer as correct. Our minds are instead like sloppily erased slates. There is still enough residual information from the past to allow reconstruction of the incompletely erased truths.

What passes for teaching is akin to the nostalgia depicted by Marcel Proust in *Remembrance of Things Past*. For Socrates, the scent of madeleine cake revives forgotten truths about Cake, Roundness, Brownness, and Nuttiness. "Learning" is akin to revisiting your childhood school. Socrates is the friendly older janitor who lets you in. He guides you through the building. When you make a wrong turn, Socrates lets memories marinate. Reoriented, you run a hand along the walls, hear the echoes of the gym, smell the cafeteria offerings. Socrates's tour lights up your memories by methodically exposing you to reminders. But the janitor is not a teacher.

Each of us is latently omniscient and therefore uneducable. All Socrates can do is elicit this knowledge from his conversational partners. At trial, Socrates explains his poverty; he never accepted tuition because teaching is impossible. In Socrates's opinion, 'Socratic teaching style' is a contradiction in terms.

Shadowing is the concrete counterpart of the abstract relationship of "participation". To understand the subtle relationship between universals and their instances, Plato builds on the relationship between objects and their shadows. The objects are widely agreed to be more real than the shadows that depend upon them. An understanding of the forms can be obtained by extrapolating the shadow relation.

Consider the relationship between the pyramids of Egypt and pyramidhood. After building an eight-level step "pyramid" south of Saqqara, Pharaoh Snefru (ca. 2600 BC) became interested in building a true pyramid—one with flat sides. His first effort at Dahshur was an equilateral pyramid with a square base. However, setting the walls at a 60-degree angle proved too steep. The building began to collapse in on itself. Snefru was forced to alter the angle for the top half of the edifice. This compromise resulted in "the Bent pyramid."

Smarting from his mistake, Snefru built the squatter Red Pyramid just to the north. Since the sides are not bent, it more closely approximates an ideal pyramid. Snefru's Red Pyramid is described as "the first true pyramid." But its sides are not as straight as those of the large pyramid his son Khufu built at Giza.

Strictly speaking, none of the pyramids of Egypt meet the mathematical definition of 'pyramid'. All physical pyramids must fall short of the ideal that guides their construction. This is inevitable because physical things are unstable. Snefru cannot perfectly align blocks. Even if Snefru could, they would not stay aligned. Bit by bit, pyramids crack and crumble. Plato's forms, in contrast, are permanent. Pyramidhood will continue after there are no more pyramids.

Plato associates reality with permanence. Since permanence comes in degrees, there are degrees of reality. The shadow of a pyramid exists for a single day and then is extinguished at sunset. The shadow is a dependent particular, depending on the object that casts it. There can be a pyramid without its shadow but there cannot be the shadow without the pyramid. The pyramid in turn depends on pyramidhood. Had Snefru not thought of the geometrical form, the pyramid would not have been erected. There can be pyramidhood without pyramids but there can be no pyramid without pyramidhood.

Plato pushed the shadow analogy further by distinguishing between lower and higher forms. Consider how a geometer deduces the surface area of a metal washer. First, he idealizes the washer as an annulus: a pair of concentric circles (fig. 9.1).

He calculates the area of the large circle and subtracts the area of the small circle. The remainder equals the surface area of washer. The annulus is a lower form governed twice over by the higher form, circlehood. Just as the washer can cast a double shadow, circlehood

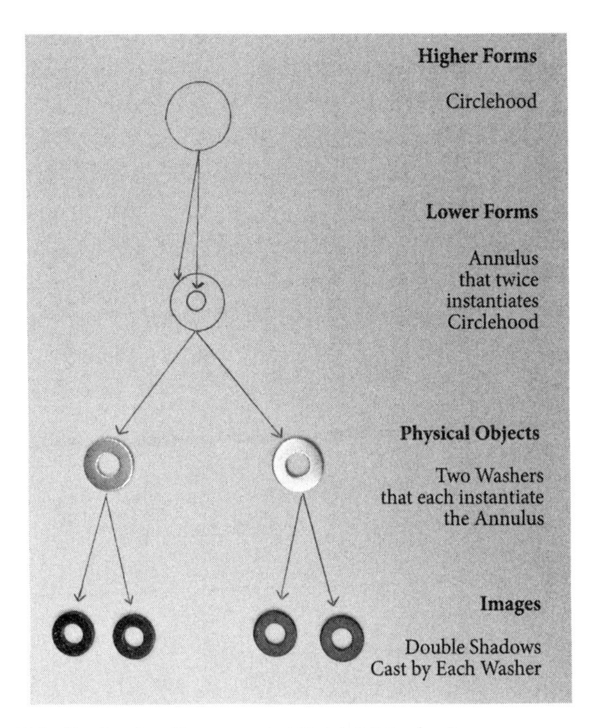

Figure 9.1 Shadowing Sequence in Plato's Metaphysics

casts the double shadow of the annulus—a circle within a circle. Higher forms govern lower forms. The relationship is one over many. The mark of being a copy is the existence of other copies. Since the copies are themselves copied, instances of circlehood cascade down the chain of being.

Copying is asymmetrical. Circlehood defines annulus. Annulus does not define circlehood. The annulus in turn defines the washer. And the washer defines its shadows.

As we gaze down the chain of being, repeated duplication makes the copies ever fainter. As we look up, things getter better defined. Definitions, however, must eventually terminate. At the top, there is at least one form that is understood without the benefit of definition. Such a form is known by rational intuition. This understanding creates premises for the deduction of further knowledge. Below this mathematical realm of knowledge by deduction is a domain of physical objects that are accessed by perception. Physical objects of the same kind are individualized by their imperfections. They are twins who grow easier to discern by their disfigurements.

Below the physical objects is the domain of imagination. In this underworld there are only images. Staring at a brightly illuminated washer causes an *afterimage*. This mental entity is presented in space but is not in space. Shadows are images that are in space but cannot be met in space. Reflections are composed of light. Shadows are constituted by an absence of light. This positive/negative duality explains why the reflection and shadow of the metal washer are located at opposite sides of a light source (fig. 9.2).

Plato characterizes higher forms as better known, as less imagistic, less hypothetical, and more logically basic than the lower forms. He is not content with any of his descriptions. Plato hints that the higher

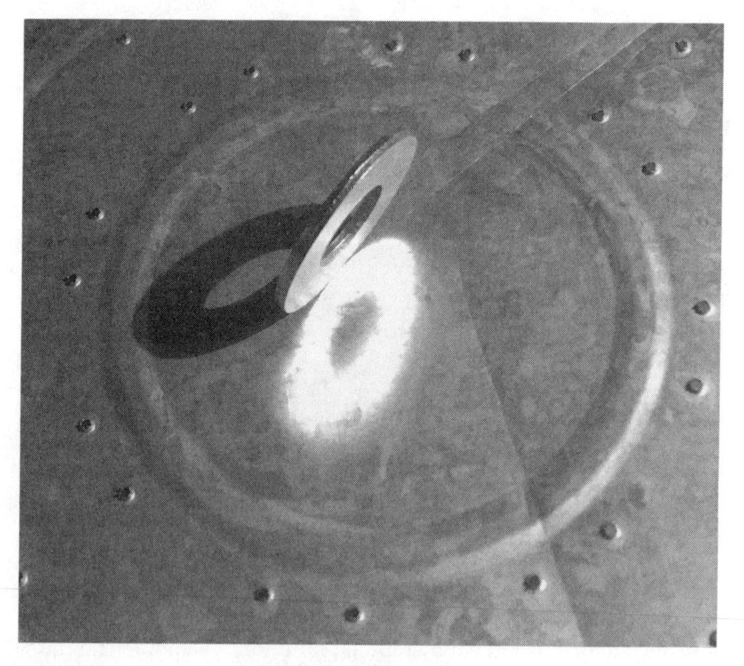

Figure 9.2 Shadow and Reflection of Metal Washer

forms are ineffable. The summit of being is at a dizzying altitude. The form of the good is "beyond being."

An Aesthetics of Absence

Shadows are the bottom-dwellers. They are the least real and therefore the least interesting of things. Shadows have the consolation of not being alone. This metaphysical slum also houses reflections, illusions, and . . . art!

There are operational and historical connections between shadows and art. A shadow is frequently pictured as an imitation of the object that casts it. In silhouette portraiture, the artist represents the sitter by tracing the outline of his shadow. Pioneers of photography described their chemical processes as "fixing the shadow": turning a transitory likeness into a permanent one.

The artist Zeuxis (ca. 424–ca. 380 BC) developed a technique for painting light and shadow. He competed with other artists to produce better imitations of the objects he depicted. The man to beat was Parrhasius, who was renowned for his trompe-l'oeil effects. In a contest, Zeuxis (ca. 424–ca. 380 BC) tried to outdo Parrhasius with a painting of a boy holding a dish of grapes. Birds flew to the picture and tried to eat the grapes. Impatient to be declared the winner, Zeuxis demanded that Parrhasius reveal his painting. Parrhasius hesitated. Impatiently, Zeuxis strode up to the painting to remove the curtain. But the "curtain" was itself painted. Zeuxis had deceived the birds but Parrhasius had deceived Zeuxis!

When the portraitist Godfrey Kneller (1646–1723) was told of this legend, he disparaged Zeuxis: "If the boy had been painted as well as the grapes, the birds would have been afraid to meddle with them" (Fadiman 1985, 333).

Plato had a more fundamental criticism of Zeuxis: an imitation of grapes cannot be as good as the grapes themselves. When one grasps the form of grapes, grapehood, one realizes that a painting of grapes is merely a shadow of a shadow.

Platonism is bad for the art business. When a friend urged the Neoplatonist Plotinus (AD 205–270) to have his portrait painted,

Plotinus refused: "It is bad enough to be condemned to drag around this image in which nature has imprisoned me. Why should I consent to the perpetuation of the image of this image?" (1991, vol. 5).

Nevertheless, historians were intrigued by the archaeologist Stephen Miller's announcement that a bust of Plato was genuine. Whereas most busts of Plato were idealized, the one in Berkeley's Hearst Museum depicts Plato with a damaged left ear—common to wrestlers struck by a right-handed opponent (fig. 9.3). 'Plato' is actually a wrestling nickname meaning "the broad shouldered one." (Plato's regular name was Aristocles—meaning "well-named.")

Resourceful artists exploit opportunities afforded by Plato's shadowy aesthetics. William Shakespeare (1564–1616) concludes *A Midsummer Night's Dream* (act 5, scene 1) with an unexpected plea for tolerance:

> If we shadows have offended,
> Think but this, and all is mended,
> That you have but slumbered here

Figure 9.3 A real statue of Plato?

> While these visions did appear.
> And this weak and idle theme,
> No more yielding but a dream.

The world of dreams is a sanctuary from morality. We do no wrong by wronging someone in a dream. We suffer no harm by being victims in a dream. So if depictions are like dreams, we have no grounds to censure actors and playwrights. They are protected by the irrelevance of fiction to reality.

Other artists use Plato's theory to rank their genre over competing genres. For instance, sculptors rank statues higher than paintings because their three-dimensionality makes them more lifelike. You can walk around them, see them from all sides, even feel their contours.

Henry Wadsworth Longfellow (1807–1882) portrays Michelangelo as trying to top both painters and sculptors:

> Ah, to build, to build!
> That is the noblest art of all the arts.
> Painting and sculpture are but images,
> Are merely shadows cast by outward things
> On stone or canvas, having in themselves
> No separate existence. Architecture,
> Existing in itself, and not in seeming
> A something it is not, surpasses them
> As substance shadow. (Longfellow 1922, 542)

Actually one of Plato's most cutting remarks applies with special force against architecture. Tall structures are especially vulnerable to the foreshortening illusion, so Greek architects distort the dimensions to create an appearance of being well proportioned. For instance, the columns of the Parthenon are tapered and tilted to make them look straight from the perspective of observers on the ground. Plato complains that the architects are guilty of copying an appearance of appearance:

> If artists were to give the true proportions of their fair works, the upper part, which is farther off, would appear to be out of proportion

> in comparison with the lower, which is nearer; and so they give up the truth in their images and make only the proportions which appear to be beautiful, disregarding the real ones. (*Sophist* 236, trans. Benjamin Jowett)

Plato exposes himself to a proof that the great chain of being has no bottom: a painting of grapes has a lower degree of reality than the grapes. A painting of a painting of grapes would have an even lower degree of reality. Lower yet would be a painting of a painting of a painting of grapes. And so on down Nagarjuna's bottomless pit.

This danger of metaphysical free fall is posed by a logical difference between shadows and paintings—and everything else Plato places at the (false) bottom, such as reflections and illusions. There can be paintings of paintings, but there cannot be shadows of shadows. Only shadows provide a stable bottom to Plato's metaphysics.

Yet Plato also needs the shadow relation to be transitive relation like the relation of being a descendant. His complex simile represented by the divided line assumes that if A is shadow of B and B is a shadow of C, then A is a shadow of C.

Transitivity is also needed in the allegory of the cave. The prisoners see a shadow of figurine of a horse rather than a shadow of the horse itself. If the shadow relation is not transitive, then the shadow on the wall does not bear the relationship to a horse that Plato alleges holds between horse and horsehood. The allegory is further confused by the fact that the prisoners can see their own shadows and perhaps the shadows of their chains. These are shadows of objects rather than shadows of representations of objects. Two types of shadows lurk in the cave.

Could Plato recast his metaphysics with the imitation relation? Imitation is transitive: if I imitate you imitating a kangaroo, I imitate a kangaroo. The problem with replacing shadows with imitations is that imitation is artificial; it depends on the aim of an imitator. When you imitate a kangaroo, you intend to resemble a kangaroo. (Since Plato was ignorant of kangaroos, he could not imitate a kangaroo.) Your shadow does not intend anything and so does not imitate you. The shadows in a shadow play are imitations because the puppeteer intends them to resemble objects.

Shadow plays can include some nonrepresentational shadows. The puppeteer sometimes makes shadows for the sake of their intrinsic beauty rather than with a view to representing anything.

Plato portrays all poetry as empirical. But a priori poetry is possible. For I have made it actual:

Ode to Pythagoras

Four
Is three
Plus one.
Three is three
Or one plus two.
Two is one
And one
Is one.

Geometrical poems are sullied with shape imagery. By keeping to numbers, I have ensured that my poem expresses literal truths and no falsehoods.

Most physical things do not imitate because imitation requires intention—or at least design. Cuckoo birds mimic the calls of their host species. Viceroy butterflies imitate poisonous monarch butterflies so that birds will not eat them. But even in this limited domain of intelligence, biologists distinguish between mimicry and fortuitous resemblance. The markings on the wings of some butterflies resemble a face. But they only resemble a face in the accidental way that the moon resembles a face.

Plato could try to universalize imitation by postulating a master designer. In the *Timaeus*, Plato speculates about a demiurge who organizes the universe out of an initial chaos of primitive elements. Perhaps the demiurge fashions all inferior things to imitate their superiors.

Plato only invokes the demiurge as a "likely story" to explain natural phenomena. Many share his doubts about whether the demiurge exists. They are also reluctant to regard imitation as the cement of the universe (rather a more general relation such as causation). As evident from the Plato's dialogues on false belief, Plato wanted to explain how representation is possible. Imitation presupposes representation and so cannot explain it.

An Epistemology of Light and Dark

As you descend Plato's chain of being, there is more and more change. Knowledge requires stability. Knowledge is therefore restricted to the forms. Once we dip to the level of physical objects, we must resort to our senses. Perception can only yield opinion. Plato uses a visual analogy to explain why vision is inferior to reason:

> And the soul is like the eye: when resting upon that on which truth and being shine, the soul perceives and understands and is radiant with intelligence; but when turned towards the twilight of becoming and perishing, then she has opinion only, and goes blinking about, and is first of one opinion and then of another, and seems to have no intelligence. (*Republic* 7.88)

The preeminence of the eye overwhelms Plato's intent to denigrate perceptual knowledge. Light is knowledge. Darkness is ignorance. Plato cannot resist portraying knowledge in visual terms. This ensures that shadows will be associated with the unknown. The study of shadows will be an impossible enterprise—an attempt to know what is intrinsically unknowable.

This skepticism about the study of shadows is embodied in the allegory of the cave. Shrewd prisoners become proficient at predicting shadows. But shadows are not objects of knowledge.

When an individual first becomes aware of the forms, he is like a prisoner in the cave who has been unchained and is free to turn his head toward the objects that cast shadows. He now sees there is more to reality than shadows. The ascent out of the cave is not smooth because he is unfamiliar with the new objects. When he reaches the surface, he is still not able to see; the strong light dazzles him.

As the freed prisoner adjusts to the upper world, he is amazed by what is revealed in the daylight. Eventually, he will be able to contemplate the source of all illumination, the sun. In the myth of the sun,

Plato compares the form of the Good with the sun. The form of the Good is the origin of everything and sustains everything. Just as sunlight illuminates everything for the eye, the form of the Good is the source of all knowledge.

If shadows are among the unknowables, why is Plato so confident in his claims about their nature? Plato can have knowledge of shadows only if there is a form for shadows. Yet Plato is reluctant to grant that there are forms for lowly things such as mud. If Plato has no knowledge of shadows, he cannot know that his metaphysics correctly uses the shadowing relation to organize the universe. Nor can he know that shadows are less real than the objects that cast them.

An Appearance-Reality Distinction for Shadows

Many people go along with Plato's characterization of shadows as illusions. But in practice, they draw an appearance-reality distinction within the realm of shadows. A dark patch that looks like a shadow can be revealed as a stain—and vice versa. People tend to call a dark patch a shadow if it resembles an object. This leads some of them to deny that undifferentiated shade is shadow.

It is true that meaningful shapes are identified more readily than meaningless ones. Meaningless shapes are interpreted as the space between objects rather than objects. Since past experience affects figure-ground organization, observers sometimes differ in whether they regard a patch as a black snake.

However shadows cannot really be dependent on our ability to classify them into some familiar shape. For recognizability varies with orientation. A dark patch that looks indeterminate from one perspective can look like a unicorn from another perspective. The shadow of the earth, night, only looks conical from the perspective an astronaut.

Viewers reject depictions of shadows that blend shadow casters with their shadows (fig. 9.4). This shadow-object hybrid is rightly regarded as an impossible object.

The Stevie Ray Vaughn Memorial includes his shadow (fig. 9.5). But this composite is not a statue of Stevie Ray Vaughn and His Shadow.

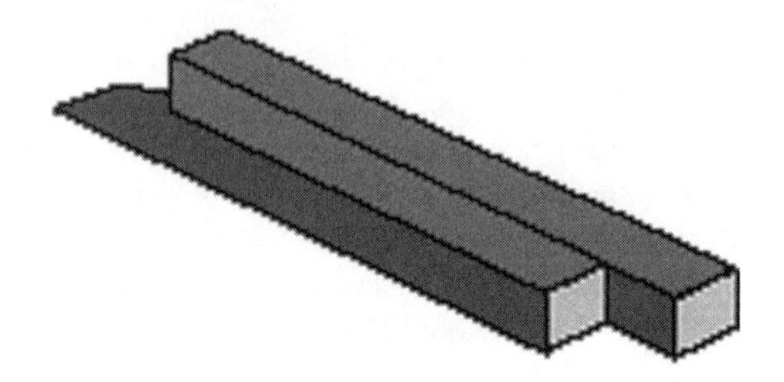

Figure 9.4 Commingled Shadow

Figure 9.5 Stevie Ray Vaughn Memorial (sculptor Ralph Helmick)

The shadow is a visual aid rather than part of the aesthetic object. Viewers typically do not notice the bronze puddle constituting the "shadow." The visual system suppresses awareness of shadows. This prevents shadows from being mistaken as objects. Shadows contain useful information about how the parts of the objects are situated with respect to each other and how the object is situated in the scene.

For instance, you compute the position of a coffee cup handle by exploiting the shadow cast by the handle against the cylindrical surface plus the shadow of the cup cast on the table plus the shadow of your approaching hand. The calculation is kept from awareness. Editing out the shadows keeps you focused on what is practically important—the physical object. The sculptor exploits the fact that a fake shadow can perform the same service as a real shadow.

A fake shadow can be covered by a white towel. Preschool children are surprised that the real shadow cannot also be covered by a white towel (Casati 2018, 158). The shadow resurfaces even darker against the white background. Authentic shadows graduate "from the Military School of Life—Whatever does not kill me makes me stronger" (Nietzsche [1889] 1998, 58).

Could Plato's prisoners draw an appearance-reality distinction within the realm of shadows? There actually are people who see only shadows. Cataracts cloud the eye so that only light and dark can be perceived. Some children have been born this way. Although they only see shadows, they see three-dimensionally.

Congenital cataract sufferers have had the advantage of tactile and discursive clues about how to interpret what they are seeing. They have felt the objects associated with the shadows, and their parents told them how to interpret what they were seeing. This encouraged the cataract sufferers to draw an appearance-reality distinction and to conceive of reality three-dimensionally. Could this be achieved without tactile and discursive clues?

This question can be answered by modifying Plato's cave. Instead of trying to deceive the prisoners, we are now trying to educate them. We labor under two constraints. We cannot educate them by their sense of touch. Nor can we narrate what they are seeing. The lessons must be taught graphically.

We should take advice from perceptual psychologists. How can they prompt the prisoners to draw the appearance-reality distinction for shadows just by providing sample shadows? What specimens would the experts choose?

One candidate is a spin-off of the Hermann grid illusion. This is a collection of shadows cast by a network of rectangular patches (fig. 9.6).

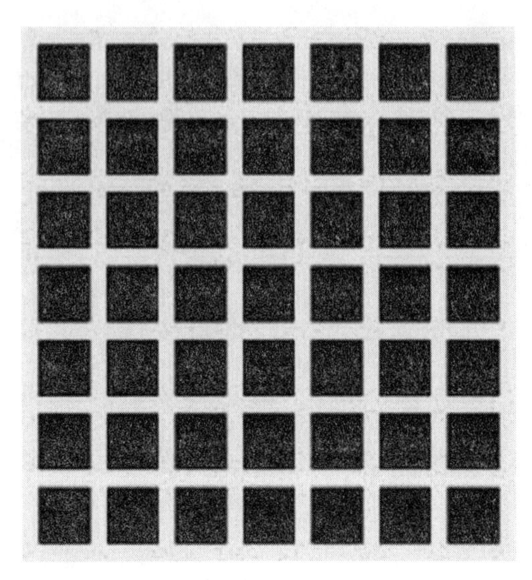

Figure 9.6 Hermann-Grid Illusion

The shadow rectangles look very black because of the contrast with the white light. At the intersections, there is more light and so less contrast. Consequently the prisoners "see" ghostly gray dots appearing intermittently. The projected gray dots are created by differences in contrast. The shadows are real things created by the contrast.

In a photographic negative of a dirty porch screen, the viewer has trouble discerning projected dots from actual dots of dirt (fig. 9.7). When the projected dot happens to coincide with a real dot, the viewer is experiencing a "veridical hallucination."[2] The content of the projected dot matches the dot but in an accidental fashion. Perception requires a lawlike connection between the content and scene. The lucky match is not enough to see the dot.

[2] Philosophers define 'illusion' as a perception that distorts a property of an object or a relationship. 'Hallucination' is reserved for distortions that introduce an object that does not exist. The Hermann grid illusion does not fit these definitions because the projected dots do not correspond to dots in the scene. Hallucinators themselves rarely conform to the definition. They suffer distortions about the nature of what is present. Rarely do they project an object such as the giant rabbit in the 1950 comedy *Harvey*. Hallucinators usually realize they are hallucinating. They are not delusional.

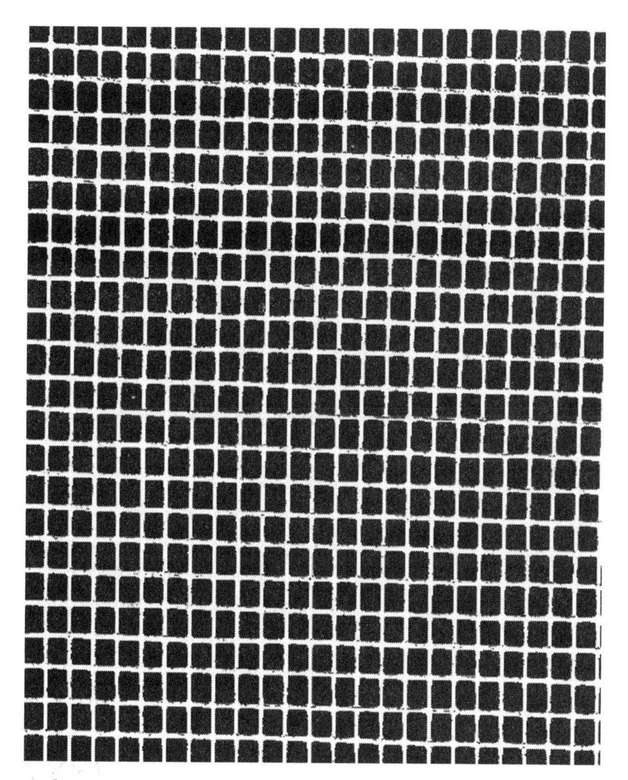

Figure 9.7 Photographic Negative of a Dirty Screen

The psychologists could next show the prisoners the shadow of a square with four Pac-Man-shaped holes. As the light shines through, we seem to get a "blacker than black" shadow within the shadow (fig. 9.8).

There really is no difference between the level of the darkness of the small "square" and the big square. But it is really just a subjective contour.

Another interesting feature of the internal "shadow" is that it seems to float above the background shadow. We tend to picture Plato's prisoners as confined to a two-dimensional world. But a two-dimensional stimulus can produce a three-dimensional representation. Wallach and O'Connell (1953) dramatically demonstrated this

Figure 9.8 Blacker than Black Shadow

with the shadow cast by a randomly bent wire. The rigid wire was mounted on a turntable so that it can rotate on its vertical axis. When the subject looks at the shadow cast by the wire, the shadow looks two-dimensional. When the wire is slowly rotated, the shadow immediately pops into a three-dimensional shape (and the viewer makes much more progress in discerning the shape of the wire). When the wire stops rotating, the shadow on the wall resumes its two-dimensional appearance. This is known as the kinetic depth effect.

Biological motion makes us connect shadows that seem unrelated when seen at rest. Psychologists show subjects six black dots that seem like a random collection. Actually, they correspond to the joint locations of a cat. This becomes apparent when the psychologists animate the display; subjects immediately connect the dots as a cat.

Shadows are more objective than afterimages. After you see the flash from a camera, you see a blob "floating" erratically about the room. Although the image is presented in space, it is not really out there. In contrast, the shadow of a tree is out there in the sunny pasture. You can measure its exact dimensions, photograph it, and even reach into the shadow to feel its coolness. But the tree's shadow is not absorbing the light or blocking the light. Shadows are absences of light.

If Plato's prisoners were shown shadows of the above sort, they would have reason to draw the appearance/reality distinction within the realm of shadows. So Plato cannot identify shadows with appearances and illusions.

Most contemporary philosophers reject Plato's view that there are degrees of reality. A thing cannot halfway exist. If it is there to have its existence measured, then it is, well, *there*. Existence is black and white. Ergo, either shadows exist or not!

I could go on and on about Plato. But I have already told you more than I know.

10
Aristotle
Potential Absence

Consider two glasses. The increasing glass is, at present, halfway toward being full (fig. 10.1). The decreasing glass is, at present, halfway toward being empty. At present, are any of the glasses half full or half empty?

Many feel compelled to answer uniformly because the two glasses are qualitatively identical; either they are both half full or both half empty. Aristotle prefers a dynamic response: the answer depends on whether you are filling or emptying. The increasing glass is half full because it is in *process* of being filled. The decreasing glass is half empty because it is in the *process* of being emptied. We should not restrict attention to the actual, present condition of the glass content. As an adjective, 'half empty' picks out a state. As a verb, 'half empty' picks out a process. Aristotle verbs adjectives.

Each thing is constituted by its history and by its possibilities. You are more than what you are at present. You are also what you *were*, what you *will* be, and what you *could* have been. Whereas the historian is restricted to what was the case, the poet ranges over possibility. Imagination liberates the poet from particularity. This universality makes poetry more philosophical than history.

Arthur Schopenhauer sharpens Aristotle's contrast to a wicked point: "The chapters of the history of nature are at bottom different only through the names and dates; the really essential content is everywhere the same" (1819, 2:442). The historian pores over the skin of humanity. The poet reaches the heart.

You should not be self-centered about change. Change requires a reciprocal relationship between the changer and the changed. A tree on a slope has a passive power to fall in the direction of the slope. Actively chopping the tree triggers a manifestation of that disposition for directed

Figure 10.1 Increasing and decreasing glasses

descent. Suppose you are not from the Taoist School of Lumberjacking. Instead of working with nature, you want the tree to fall in the opposite direction of the slope. Now you must alter the tree's latent trajectory by notching the trunk. Actively chopping puts this hinge into operation.

"Be all you can be" is impossible. Your potentialities are not co-realizable. As some potentialities are actualized, others become defunct. "Every man is born as many men and dies as a single one." Martin Heidegger *could* have said that. Actually he did not.

Time-wise, potentiality precedes actuality—a man must first be a boy. Definition-wise, actuality precedes potentiality. For the actual must be cited in characterizing the potential. To be a boy is to be capable of developing into a man. To be visible is to be capable of being seen. 'Buildable' means "capable of being built."

Disputes over whether the glass is half full or half empty are equivocal when different goals are assumed. If the different goals are common knowledge between the disputants, then the question only makes sense as a disagreement as to which goals ought to be pursued.

Relative Absences

If nobody is in a completely stocked library, is the library empty? We answer by relativizing: the library is empty *of people* but full *of books*. Most sophistical arguments about absences equivocate between different relata.

> You are not in Stagira.
> Hence, you are somewhere else.
> If you are somewhere else, you are not here.
> Therefore, you are not here.

The solution to the absence paradox is that 'here' is relative to a place. The conclusion is false when 'here' is relativized to the place where the conclusion is asserted, but it is true relative to any of the places covered by 'somewhere else'. The relativity of absences allows Aristotle to distinguish absences instead of lumping them into a tangled ball of nothingness.

There are many more ways of not being than being. For there is only one actual outcome yet many potential outcomes. This explains why positive descriptions are more specific than negative descriptions, why success is more difficult than failure, and why virtue is narrower than vice (*Nicomachean Ethics*, Book 2).

Being is not a monolith. Still less is nonbeing. Aristotle allows that omissions can be causes. His example is a shipwreck caused by the absence of the pilot. But causation by absence must always be made within the framework of being. 'Absence' is a kind of placeholder for something positive.

Actual Absence versus Potential Absence

Aristotle contradicts Anaxagoras. There are many relative absences—just no absolute absences. Aristotle would regard Anaxagoras's answer to "Is the glass half empty or half full?" (*Neither; nothing is completely empty, nothing is completely full*) as closer to the truth than Leucippus's answer (*Both; half is water and half is void*). By denying that the glass is half empty, Anaxagoras correctly implies that there is no void.

However, Aristotle would reject Anaxagoras's ban on pure specimens. The water in the glass could be pure water.

Anaxagoras thought that X can change into Y only if everything comes premixed. Aristotle's alternative is that mixing can occur by activating the potentials of ingredients. Adding a spoon of sugar to the water leads to a genuine mixture, sugar-water, because the ingredients survive and contribute to the new, uniform whole.

Potential Emptiness

Aristotle's only concession to Leucippus is that emptiness can be approximated. Air more closely approximates the void than water. But this does not mean that the void is possible. Figures can come closer and closer to being a round square (fig. 10.2). But these approximations do not show that the round square could be actual.

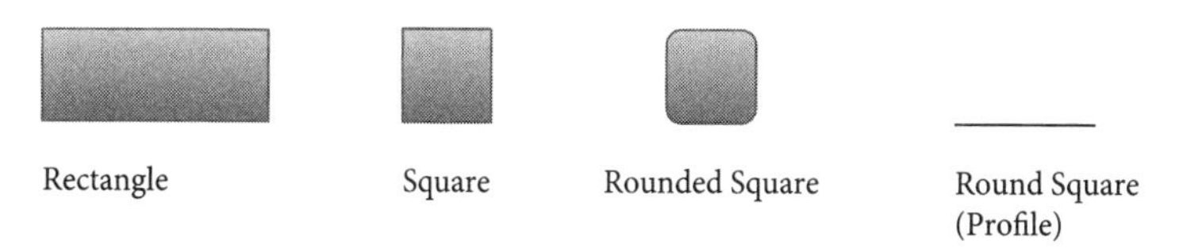

Figure 10.2 Approximating the round square

An object can become more rarified because substances are elastic. They are not fixed in size like atoms. No interstitial vacuums are needed because matter s-t-r-e-t-c-h-e-s. A marble tile swells with the summer heat. As density decreases, there is less matter per unit area. So space is occupied less intensely. But the space is never unoccupied. Rarification can never be total.

The void resembles infinity: always potential, never actual. If you keep bisecting a line, smaller and smaller magnitudes result. But the sequence does not end in a smallest magnitude.

The existence of a smallest magnitude would solve Zeno's bisection paradox. According to Zeno, Leucippus cannot exit the room because he would have to travel halfway, then half the remaining distance, that half that smaller distance, and so on. Leucippus can reply the he need only cross finitely many of these minimal units.

Aristotle avoids commitment to a smallest distance. He solves the bisection paradox by denying that there is any actual infinite. The only kind of infinity is open-endedness. However high you count, you could have counted higher. But you still only count finitely high.

On the Offensive with Motion

Leucippus's principal motive for postulating the void is to explain motion. Aristotle objects that this backfires: the void would actually make motion impossible.

Plato's atoms move as members of a queue; the last member moves into the position vacated by the penultimate member who in turn assumes the position of his predecessor.

Leucippus rejects this succession model because any movement would require massive coordination. It is imperative that Leucippus not repeat the problem by explaining the motion as displacement of the void.

If you put a gold cube in a glass of water, the level of the water rises because the cube displaces an amount of water equal to its own volume. A cube of water and a cube of gold cannot exist in the same place at the same time. But if the cube does not displace the void and the void continued to exist, there would be two things in same place at the same time. And if it is possible for one void to coincide with the gold cube, why not two voids or any number? (*Physics*, 4.8.216b8–11).

This is the first recorded application of principle of lightness to *absences*. Indian philosophers had long applied the principle to beings. But never to nonbeings. Aristotle minimizes how many absences there are. He scrimps on the *kinds* of absences permitted (allowing omissions but forbidding voids) and the number of absences of that kind. For instance, Aristotle prefers to postulate few omissions than many omissions. In the case of the shipwreck, Aristotle believes there is an absence of the captain piloting a ship but not an absence of the captain's mother piloting the ship. Those wishing to improve the safety of ships record a limited number of near shipwrecks. Postulating too many near misses will overwhelm our limited attention.

An economist may fail to detect any frugality in minimizing absences. Suppose you have 10 unused credit cards and so are billed zero dollars each by the 10 creditors. I have one unused credit card and am billed zero dollars by my single creditor. The economist judges us as equally parsimonious.

Aristotle would dismiss the economist as innumerate. Zero is not a number. Since Aristotle had no sons, he did not have a number of sons. Aristotle did have a daughter. But he would still deny he had a number of daughters. One is not a number. The first number is two. Common sense concurs with Aristotle: Aristotle did not have a number of daughters; he had one!

From Absence of Reason to Reason for Absence

Given Leucippus's aphorism "Nature does nothing in vain," there is a reason for everything. Arbitrariness is impossible. So it is equally true that Nature *omits* nothing in vain. The pair of tiny holes in a human jaw seem like oversights to contemplators of skulls. But the mandibular foramen are canals for nerves that connect face and neck muscles.

Aristotle is ambivalent about arbitrariness. On the one hand, he thinks a perfectly uniform hair can snap even though there is no reason for the hair to snap at one location rather than another. On the other hand, Aristotle praises Anaximander's explanation for the stability of the earth: there is no more reason for the earth to move one way rather than another. Absence of reason (*for* motion) is reason for absence (*of* motion). This ban on arbitrary motion applies to any object in the void. After all, the void is completely homogenous. There is no more reason to move in one direction than another. So any object in a void would be at rest.

Leucippus would reply that the atoms have always been in motion. He is independently committed to the principle that the amount of motion does not decrease. For if the atoms had any tendency to slow down, they would have already exhausted all motion. After all, the past is infinite. Atomists must conclude that there has always been motion and there will always be motion. The universe is a perpetual motion machine.

Ordinary observation runs contrary to perpetual motion. Although *celestial* objects, such as planets, constantly move, terrestrial objects slow down and stop. Even if the atoms began in motion, the terrestrial atoms would have come to rest by now.

Grant the atomists perpetual motion. They would still be falsely predicting atoms flit about with unlimited speed. For the vacuum offers no resistance. The Islamic Aristotelian Avicenna (980–1037) added the objection that a finite cause (nudging a stationary atom) cannot have an infinite effect.

Grant the atomist unlimited speed. There would still be problem of the object being in more than one place at the same time. For it passes through intermediate positions but not at successive times. The object is simultaneously at all points along its trajectory!

Is multilocation impossible? In 1940, John Wheeler thought the best explanation of why all electrons are qualitatively identical is that they are numerically identical; there is just one, multilocated electron. If Leucippus had thought all atoms were qualitatively identical, he might have come closer to Parmenides: the universe is composed of a single atom. The multilocated atom would out-economize Claud Mellan's single-line portrait of Jesus, which spirals out from the tip of Jesus's nose (fig. 10.3).

Aristotle's final objection to the atomist account of motion is that the speed of objects moving in a vacuum would be absurdly *uniform*.

Figure 10.3 Sudarium of Saint Veronica, engraving by Claude Mellan, 1649

If a heavy object were released at the same time as a light object, then they would reach the ground at exactly the same time.

The Christian experimentalist John Philoponus (about 490 to 570) reports that this predicted uniformity comes close to being realized in normal conditions: the heavier object only slightly exceeds the lighter. Philoponus's explanation: light objects are *less* affected by air than heavy objects. Observe how tall trees are toppled by gusts that leave little trees standing. If the objects fell in a vacuum, the heavier object would fall much faster than the lighter object!

Galileo replicated Philoponus's experiment (legendarily from the leaning Tower of Pisa). However, Galileo discreetly sided with Aristotle's conditional prediction of what would happen if the two were dropped in a vacuum. According to Galileo, lighter objects are *more* affected by air. Observe how tall trees stand still in a breeze that bends saplings.

Aristotle's thought experiment eventually became an actual experiment. Apollo 15 commander David Scott dropped a feather and a hammer on the moon:

> During the final minutes of the third extravehicular activity, a short demonstration experiment was conducted. A heavy object (a 1.32-kg aluminum geological hammer) and a light object (a 0.03-kg falcon feather) were released simultaneously from approximately the same height (approximately 1.6 m) and were allowed to fall to the surface. Within the accuracy of the simultaneous release, the objects were observed to undergo the same acceleration and strike the lunar surface simultaneously, which was a result predicted by well-established theory, but a result nonetheless reassuring considering both the number of viewers that witnessed the experiment and the fact that the homeward journey was based critically on the validity of the particular theory being tested. (Allen 1972, 2–11)

Twentieth-century space travel resurrected many objections to movement in empty space. In a 1920 editorial, the *New York Times* attacked Robert Goddard's claim that a rocket would work in space:

> That Professor Goddard, with his "chair" in Clark College and the countenancing of the Smithsonian Institution, does not know the relation of action to reaction, and of the need to have something better than a vacuum against which to react—to say that would be absurd.

> Of course he only seems to lack the knowledge ladled out daily in high schools.

In 1969, days before Apollo 11 landed on the moon, the *New York Times* published a retraction: "Further investigation and experimentation have confirmed the findings of Isaac Newton in the 17th century, and it is now definitely established that a rocket can function in a vacuum as well as in an atmosphere. . . . The *Times* regrets the error." However, the retraction "corrects" the wrong mistake. The rocket moves forward by pushing against particles of its own exhaust. The real mistake was to assume the rocket was in a vacuum. If there were only a single indivisible atom, then it would face the impossibility described in the 1920 editorial.

No Extracosmic Void

If there is a void, it is a space that lacks an actual body but which could be occupied by a body. However, no body can exist beyond the heavens. Therefore, there is no void space there. We live in a finite, spherical universe beyond which there is "neither place, nor void, nor time."

Despite Aristotle's judgment of impossibility, the extracosmic void became a sanctuary for those who wished to answer "Where is God?" To answer "nowhere" smacks of atheism. To answer "everywhere" also smacks of atheism. For there is no contrast between God and Nature. However, if God were in any local place, he would be limited. Is there any place for God? Yes, if God is in the cosmic void, he encompasses the world without being part of it.

The appeal of the cosmic void as a sanctuary continued for twentieth-century atheists. On July 18, 1969, American astronauts were scheduled to land on the moon in two days. The author of the visionary science fiction classic *2001: A Space Odyssey*, Arthur C. Clarke, was asked for his dreams and predictions:

> There is always the fear, of course, that men will carry the curse of their animosities into space. But it is more likely that in the long run, those who go out to the stars will leave behind the barriers of nation

> and race that divide them now. There is a hopeful symbolism in the fact that flags will not wave in a vacuum; our present tribal conflicts cannot be sustained in the hostile environment of space. (1969, 31)

Aristotle does not explain what happens to a spear that is thrown at the edge of the universe. Others ventured answers on behalf of Aristotle. In the fourteenth century, Jean Buridan proposed that the exiting spear would change the extent of space. This solution exploits the principle that there is space wherever there is body. Ergo, a spear-shaped envelope of space would grow to exactly accommodate the spear. Trying to go past the limit of space is like trying to wrap your fingers around a coin trapped between sofa cushions. Your fingers open new space into which the coin descends. Trying to cross the boundary between space and nonspace is self-defeating because the size of space expands to accommodate its contents.

In the sixteenth century, Giordano Bruno suggested that the spear passes out of existence. This response violates the principle that matter is conserved. For the spear would turn into nothing, thereby reducing the overall amount of matter in the universe.

A final proposal is that the questioner commits the fallacy of composition: each thing in the universe has a place; therefore, the universe has a place. For Aristotle, the concept of place does not apply to the universe itself.

Does Nature Abhor a Vacuum?

If you hold your finger over the top of a flexible tube, you can pick up water. The water will forgo its natural downward motion because that would create a vacuum. More dramatically, the water will even *reverse* its natural motion. If you bend the tube beneath the waterline of the container, the water will flow *up* the tube and out of the container.

The siphon became incorporated into the "Pythagorean cup." The cup enforces the Aristotelian ideal moderation. If the imbiber pours too much wine into the cup, a siphon is activated. All the wine is emptied through a hole at the bottom of the cup. If only a moderate amount of wine is poured, the cup performs normally (fig. 10.4).

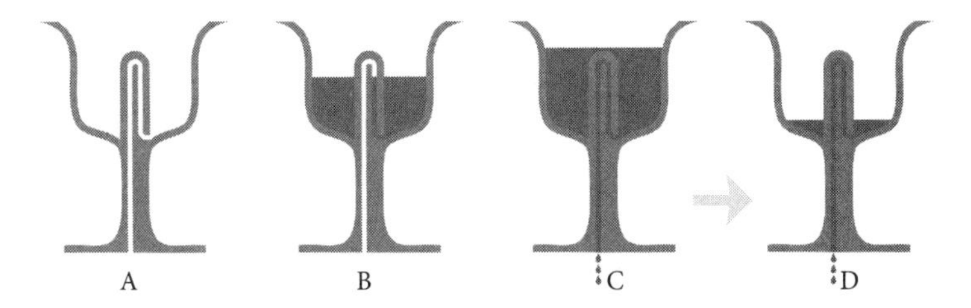

Figure 10.4 Pythagorean cup drawing by Nevit Dilmen

Modern toilets work like the Pythagorean cup. There are two reasons for this: number one and number two.

The mechanism is also the leading explanation of the intermittent stream described by Pliny the Younger in the last letter of book IV of *Letters*, 4.30. He is famous for his accurate description of the Vesuvius eruption of AD 79 that killed his uncle Pliny the Elder. Pliny the Younger also deserves credit for accurately timing the rhythmic water flow at the Villa Pliniana spring on Lake Como in Italy. Geologists believe there is an underground water source that fills a reservoir until a siphon forms. The water streams out to feed the stream until the water column is broken. The water stops. Pliny's schedule for the cycle remains accurate.

There is a myth that Aristotle explained the siphon with the principle "Nature abhors a vacuum." Christian disciples of Aristotle coined that slogan much later. Aristotle himself would have found its anthropocentrism excessive. Emphasizing different words in the slogan brings out its anti-Aristotelian aspects.

First, "*Nature* abhors a vacuum" suggests that the vacuum might be formed by an enterprising individual with different tastes than Nature. Consider the engineer Hero of Alexandria. He believed that vacuums are unnatural only in that they do not form spontaneously. A vacuum lover might create a vacuum by plugging a bellows and separating the handle (fig. 10.5).

Aristotle regards all such recipes as misconceived. Infinite force would be needed to create a vacuum. Since infinity is always potential, never actual, the vacuum is always potential, never actual.

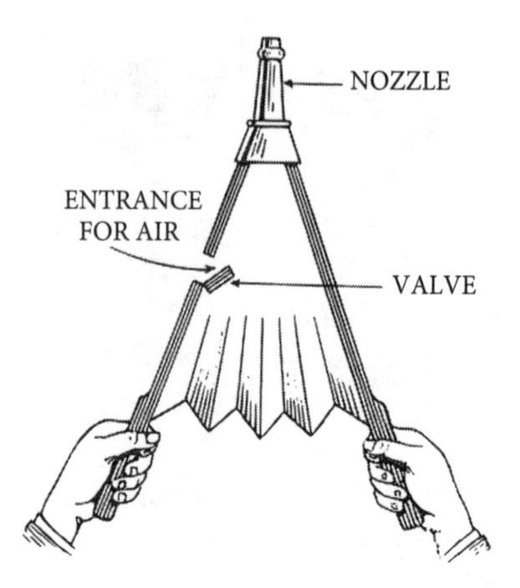

Figure 10.5 Bellows drawing by Pearson Scott Foresman

Second, "Nature *abhors* a vacuum" suggests the operation of a desire. Acting on a desire takes time. Nature would need to spot the incipient vacuum, recognize the threat to her wishes, and then abort the embryonic void. But even a brief vacuum refutes plenism (Aristotle's doctrine that nature is full).

Third, "Nature abhors a *vacuum*" suggests that the vacuum can have effects such as provoking Nature's wrath. But Aristotle believes that only positive things can be causes. (Omissions figure in causal explanations but only as a shorthand for positive causes.)

Given the impossibility of a vacuum, any abhorrence of the vacuum would manifest ignorance *or* irrationality. Nature does not waste energy guarding against impossibilities.

Aristotle would also be puzzled about why Nature would oppose a vacuum. What would Nature have against the void? What did the void ever do to Nature? The void is not an appropriate target of animosity.

Theologians later developed an indirect rationale for Nature's aversion to a void at any scale. If God exists, then the actual world is the best of all possible worlds. The best of all possible worlds seizes every opportunity to fill space with an existing thing. A void at any scale would be a waste. All of space is therefore full of existing things.

This theological rationale for opposing the vacuum was unavailable to Aristotle. He did not believe in a personal God. Admittedly, Aristotle postulates a Prime Mover who is the ultimate lure that attracts all things. However, the Prime Mover does not abhor anything. The Prime Mover is engaged in only the best activity—which is thinking. And the Prime Mover only thinks about the best topic—Itself. Thus the single-minded Prime Mover is completely self-absorbed, oblivious to the commotion it excites in the rest of the universe.

If Aristotle were going to anthropomorphize Nature, then he ought not to anthropomorphize Nature as ignorant or stupid or wasteful. For then Aristotle could no longer appeal to Nature's efficiency in explaining why water runs downhill rather than up and in a straight rather than a crooked line on a flat surface.

Anthropomorphizing Nature also confers too much flexibility. If Nature abhors a vacuum, why should Nature suffer the mischief of water traveling up the siphon? Nature should just break the siphon. The ancients believed that Nature did exactly this when pipes break from freezing. The water in the pipe threatens to create a vacuum by contracting into a solid. Nature cracks the pipe.

These thinkers did not realize that freezing water expands rather than contracts. Freezing a glass half full of water makes it more than half full, not less than half full. The popularity of the idea that "freezing water contracts" among the ancients is puzzling because of the accessibility of counterevidence. Water regularly freezes in Greece. The Greeks and especially the medieval Europeans failed to heed readily available evidence that freezing water expands. The phase structure of water was only discovered in the eighteenth century. Joseph Black (1728–1799) became puzzled why snow melts gradually rather than all at once.

Nowadays we are more open to relativizing 'nature' to a particular environment. What is natural on earth is not natural on the moon. When the Apollo 11 astronauts landed on the moon, they could not pull open the door. Too much air remained in the lunar lander. On the moon, nature abhors the absence of a vacuum!

After the astronauts removed the air, pressure was equalized, so they could exit. They were careful to leave the door ajar.

11

Lucretius

Your Future Infinite Absence

The Roman poet and philosopher Lucretius Carus (99–55 BC) was among the disciples of Epicurus. Master Epicurus had solved the major philosophical problems of metaphysics (through atomism) and ethics (through hedonism). Just as the nineteenth-century physicists believed that the future belonged to engineers, the Epicureans believed that the future belonged to therapists.

Lucretius did not seek to outdo the Greeks by innovating at the theoretic level. As a practical Roman, he works with Epicurus's *On Nature* as his sole philosophical source (Sedley 1998). Lucretius's originality lies in making poetry a therapeutic tool. He compares his poems to the honey that is smeared on the rim of a cup containing bitter medicine.

Epicurus the Master

Lucretius's book-length poem, *De Rerum Natura* ("On the Nature of Things"), follows the structure of Epicurus's scholarly presentation of his philosophy. Each begins with atoms and the void. Whereas Democritus's void is a kind of negative substance, Epicurus treats voids as empty spaces. Voids cannot move. They are destroyed when an object moves into them. When the object moves out, a new space is created. Given an infinite past, every void is destroyed. All present voids have finite ages. They will all be destroyed. Only atoms are immortal.

Lucretius emulates Epicurus's procedure of scaling up from atoms to middle-sized objects. We number among the combinations of atoms.

Our psychology emerges step by step from physical combinations. Epicurus took these mental aspects more seriously than Democritus—who is inclined to demote beliefs and desires to secondary phenomena.

Democritus deflates educational aspirations by saying, "In reality we know nothing; for the truth is in an abyss" (Diogenes Laertius IX, 72). Epicurus, in contrast, is optimistic about our ability to gain knowledge of the world. Knowledge is rather easy—as is the acquisition of happiness.

Mental phenomena at this larger scale provide clues as to what is happening at the invisible, smaller scale. In particular, introspection reveals the existence of voluntary action. This shows that the atoms cannot be entirely deterministic (contrary to Democritus). They must occasionally move by chance.

These swerves also explain how atoms start colliding with each. Atoms fall in the sense that they move parallel to each other. (Aristotle defined falling as movement to the center of the universe, but the atomists deny that there is a center. The universe is infinite and therefore lacks outer boundaries to define a center.) If there were no swerves, atoms would fall like rain in a dead calm, never interacting to make any compound objects.

Thanks to elemental freedom, there are grains of sand, sandstone, and sandstone statues. We might end the sequence with our world. But Lucretius insists our world is just one among many. The universe is infinite. There must be endlessly many worlds. Just as the sheer number of grains speaks against any grain being of privileged significance, the sheer number of worlds reduce each other to insignificance. Our world is just a big combination of atoms. It too will perish.

Given our world's lack of privilege, we should reject theological explanations of its peculiarities. You do not appeal to the gods to explain a crack in a pebble. By parity of reason, you should not appeal to the gods to explain a canyon in the earth. If thunderbolts are weapons hurled by Zeus, then why does the god waste so many strikes on deserted areas? And when Zeus does hit something, why is it so frequently his own temple?

Lucretius's antireligious motifs ensured literary suppression and character assassination. After the Christians took control of Rome, Saint Jerome (ca. 347–420) reports that Lucretius was driven mad by a love potion and committed suicide. The merits of Lucretius's poem were explained away as Cicero's editorial refinements. *De Rerum Natura* was hardly read and passed out of circulation.

Fear of Fear of Death

As with any other complex thing, you must eventually decompose. Your death is not bad for you. It will not happen in your lifetime! *Fear* of death is bad for you. This painful state can be subdued with knowledge. The purpose of Lucretius's poem is to supply this cognitive therapy.

As a first step, Lucretius lays out a metaphysics in which the afterlife is impossible. Since death is your permanent destruction, you will not suffer the hell of Tantalus, Tityus, or Sisyphus.

Incomplete application of this insight risks an irrational residue of apprehension. If you picture your corpse with a residue of life, then you will regard yourself as being sufficiently conscious to suffer in a state of impotent paralysis! Pseudo-thanatos, false diagnosis of death, was a rational fear. The clinician Katherine Hall (2019) conjectures that the 32 year old Alexander the Great was declared dead six days prematurely because his physicians did not know about Guillain-Barre syndrome. They thought Alexander's lack of bodily decay was due to his divinity. As befitting the son of Zeus, Alexander's body perfumed his clothes with an agreeable fragrance (Plutarch 1914, 465). Bishop George Berkeley did not want his reputation as an immaterialist to foster pseudo-thanatos. His will specified his body be "kept five days above ground, . . . even till it grow offensive by the cadaverous smell" (1948, 8:381)

Since nobody alive has ever been dead, we become apprehensive of this unprecedented state. To make death more familiar, the Greeks compare death to sleep. We do not fear the loss of consciousness that comes with slumber. Death, as permanent loss of consciousness, is just more of what we do not fear.

Lucretius's second precedent is your own "experience" with *past* nonexistence. You have already "been" through prelife nonexistence! How bad was it? Not bad at all. Could your future nonexistence be any worse? There is no relevant difference. So you should not expect to suffer after death:

> Look back:
> Nothing to us was all fore-passèd eld
> Of time the eternal, ere we had a birth.

> And nature holds this like a mirror up
> Of time-to-be when we are dead and gone.
> And what is there so horrible appears?
> Now what is there so sad about it all?
> Is't not serener far than any sleep? (Lucretius 1916, book III, ll. 968–77)

Lucretius merely concludes that your future nonexistence *will* not bad be for you. Is your future nonexistence bad for you *now*?

Lucretius answers no by borrowing Epicurus's dating argument. If some event harms you, then it harms you at some specific time. Death cannot harm you *before* it occurs (because that would be reverse causation) or *while* it occurs (because as soon as it occurs, you cease to exist) or *after* it occurs (because you are no longer around to be affected by it).

Swamped by Infinity?

Toward the end of *On the Nature of Things*, Lucretius argues that prolonging a life cannot make it better. He reasons that any finite improvements would be swamped by the infinite period of nonexistence.

The Greeks were more geometrical than arithmetical. Their metaphors for a good life are spatial: balance, symmetry, and proportion. Just as being longer does not make a story better, a life is not improved merely by lengthening it. Extra life is good only insofar as it completes your story. Consider the artist Thomas Hart Benton. He was described as having a perfect death because he had, at 85, just completed his final mural, *The Sources of Country Music*, for the Country Music Hall of Fame in Nashville, Tennessee. Benton contrasts with Franz Schubert. Schubert's death in 1828 at age 31, prevented him from finishing his Symphony no. 8 in B minor, commonly known as the "Unfinished Symphony."

When Jeremy Bentham revived hedonism, he only counted the quantitative aspects of pleasures as morally important. Given that the amount of pleasure is equal, a truncated life is just as good as one bearing the aesthetic merit of narrative completeness.

Bentham's quantitative approach implies that the worst time to die is at the beginning. Given that life begins at conception and that the balance of pleasure over pain grows, abortion would be the worst deprivation.

Offsetting this antiabortion theorem is the utilitarian indifference to how utility is distributed. A fetus is just an empty vessel into which value is poured. Abortion is easily compensated by the creation of another empty vessel. The governing concept is *opportunity cost.*

What applies to fetuses applies to adults. Their deaths can be justified by the creation of other people—especially people who are more easily pleased. This is the premise for "utopian" novels such as *Brave New World*, in which eugenicists shift emphasis from making people happy to making happy people.

Fear of death is based on prudence rather than morality; I am not consoled by the prospect that my death would cause *another* person to live a more pleasurable life than mine. Even when one willingly sacrifices one's life for another, it is a *sacrifice*.

From the prudential perspective, I want my life to go on indefinitely. At any stage of life, it is rational to want to continue. A desire to stop living is therefore a sign of irrationality. Accordingly, we paternalistically intervene against suicide attempts. If someone's life is going well, then a decision to commit suicide should be thwarted.

To put the point more grandly; it is never rational to want to die (given that one's life is going well), so it is rationally *mandatory* to want immortality.

Why stop here? Once you reflect on the symmetry between pre-vital nonexistence and posthumous nonexistence, it becomes preferable to have an earlier start. Or better yet, to have always been around. The insatiability of this desire for life gives it the infinite scale needed to overcome Lucretius's swamping argument.

Lucretius may have found risible a desire to have begun life earlier. But he also found it risible that Rome had an antipode at which day and night are reversed and the animals move upside down.

Reversing the Symmetry

The repertoire of the musical humorist Peter Schickele includes the "Unbegun Symphony." Schickele regretfully explains he was "born too late to write the first two movements."

From a logical point of view, you might conclude that the symmetry between pre-vital nonexistence and posthumous nonexistence shows you should be horrified by your pre-vital nonexistence.

The Romans themselves laughed off any desire to have started existence earlier or having existed earlier. Seneca writes,

> Doesn't the person who wept because he had not been alive a thousand years ago seem to you an utter fool? Equally foolish is he who weeps because he will not be alive a thousand years' time. These two are the same: you will not be, nor were you. Neither time belongs to you. (*Epistulae Morales* 77.11, trans. Warren 2004, 70)

Seneca extends fatalism about the past (the past cannot be changed) to fatalism about the future. We should serenely accept what is inevitable because resistance is futile. This wisdom about the inalterability of the past explains our tranquil resignation to pre-life nonexistence. The fatalist goes on to infer it warrants comparable composure toward our posthumous nonexistence.

According to the fatalist, we are necessary beings. Our existence could be neither longer nor shorter. Nor could any nonexistent person become existent. Since deprivation requires the possibility of an alternative, there are no winners or losers in the "lottery" of existence. All individuals are necessary existents. All nonexistent "individuals" are necessarily nonexistent.

The fatalist is no gladder of his existence than he is glad of the existence of the number 10. Both are necessary beings. The fatalist does not experience a sense of being thrown into existence. There is no vertigo of contingency. Only serenity.

V

DIVINE NOTHING

12

Saint Katherine of Alexandria

The Absence of Nonexistent Women Philosophers

My policy has been to compensate for discrimination against nonexistent philosophers by highlighting their contributions. Awkwardly, all of the nonexistent philosophers so far discussed have been men.

For relief, I appeal to the patron saint of philosophers, Katherine of Alexandria. She is the Christian answer to the pagan's Hypatia of Alexandria—a female philosopher ripped apart by a Christian mob. In third- and fourth-century Alexandria, women and philosophers were supposed to be untouchable. Murder of a woman philosopher was double trouble. Christian authorities failed to address the outrage. This elevated Hypatia to the status of a pagan martyr. Christians needed a countermartyr to balance the scale of justice.

Katherine is first mentioned in the ninth century but is reported to have lived in the fourth century. During the fourth century, faith was more important than erudition. Christians portrayed scholarship as hubris. By the ninth century, however, Christians had become more anxious about their intellectual credentials. Their ancient martyrs were re-educated.

Bloodline and beauty also became far more important than suggested by the New Testament. Katherine was described as the lovely daughter of King Costus of Alexandria and Princess Sabinella of Samaria. Katherine was well versed in all the arts and sciences. Such education would have been nearly miraculous for a fourth-century woman. However, a small number of women did study at the Museum of Alexandria. Accordingly, Katherine converted to Christianity in the course of her studies.

On one account, the seventeen-year-old Katherine went to Armenia to search for a future husband who could match her beauty and learning. A hermit named Adryan instructed her to pray that she be allowed to see Mary's son. The next day she learned that Christ rejected her as ugly because of her vanity. Chastened and catechized, Katherine again sought another vision. Christ now accepted her. He presented a ring to signify their marriage.

The next year Emperor Maximinus decreed that all Roman citizens must sacrifice to the gods. Alarmed by how many Christians offered the sacrifice to escape persecution, Katherine confronted the emperor. Overwhelmed by her logic, the emperor summoned 50 philosophers to rebut her. Outnumbered but not out-syllogized, Katherine proved that the pagan Greek poets and philosophers support Christ. The 50 philosophers converted.

Emperor Maximinus was a sore loser. He had the converted philosophers burned. "Light a fire for a man and he'll be warm for a night, light the man on fire and he'll be warm for the rest of his life."[1] The massacre of the 50 philosophers was especially vindicative because the converts burned *without the benefit of baptism*. This meant the roasted philosophers would be reroasted in hell. Some merciful medieval painters show an angel baptizing the philosophers in the midst of their immolation, perhaps the origin of the expression "baptism by fire."

What to do with Katherine? In a 180-degree emotional turn, Emperor Maximinus proposed marriage. As a bride of Christ, Katherine refused. In a second 180-degree pivot, the jilted Maximinus ordered Katherine to be killed on a spiked wheel.

Miraculously, the wheel disintegrated before harming Katherine. Unmerciful medieval painters depict bystanders being killed by splinters.

With waning imagination, Maximinus next ordered that Katherine be beaten and imprisoned without food. However, she was saved from starvation by a dove that delivered food through the bars of her cell.

Curiosity piqued, the empress and an army officer visited Katherine. When the visitors and her guards witnessed Katherine being attended

[1] This is a 1990s subversion of a 1940s proverb that is understandably misattributed to Jesus: "Give a man a fish, and you feed him for a day. Teach a man to fish, and you feed him for a lifetime." The twelve apostles are fishermen. The gospel teems with stories about fish and fishermen. In the second century, the ichthys symbol for Christianity began as graffiti but graduated into artwork.

by angels, all converted. The emperor slew this second group of fresh Christians.

Tiring of these collateral casualties, the emperor ordered that Katherine be decapitated. As her head separated, milk rather than blood flowed from her neck.

At this time, around 305, angels transported her body from Alexandria to Mount Sinai, where it lay hidden, until the ninth century. The monastery, built on the site of Moses's burning bush, took her name after the discovery of her body. The monks still venerate her relics.

Greek Orthodox and Catholic hagiographers concede there are eyebrow-raising parallels with Hypatia. What is the evidence that she existed? Her biographers appeal to Eusebius. In the fourteenth chapter of book VIII of his church history, Eusebius describes an Alexandrian beauty who might be the basis for the legends. She had education, wealth, and the support of a good family. When she resisted the advances of Emperor Maxentius, he exiled her and confiscated her property. However, Eusebius does not describe her being martyred.

There is evidence that Saint Apollonia was based on a real woman (Lewis 2000, 46). She was an elderly deaconess in Alexandria who threw herself into a fire rather than recite pagan prayers. As the tale is told and retold, Saint Apollonia becomes younger and her demise becomes more lurid. By the late Middle Ages, Apollonia is a beautiful young virtuous Christian who is ingeniously tortured by lascivious pagans.

There is a four-hundred-year gap between Katherine's alleged martyrdom and there being any literary or pictorial reference to her. All present versions of the legend can be traced back to an account written by Simeon Metaphrastes written between 960 and 964. The threads of the story are too frayed to connect with anyone.

Saint Katherine teaches us lessons about celebrity nonexistents. Just as the supernatural is modeled on the natural, the nonexistent is modeled on the existent. We lack the imagination to stray far from what is familiar. So we make a few dramatic alterations and let habitual presuppositions fill in the rest of the story by default. As the number of women philosophers increase, so will increase the number of nonexistent women philosophers.

13

Augustine

The Evil of Absence Is an Absence of Evil

Augustine's metaphysics resembles the Temple of Concordia—a Greek temple that was converted to a basilica in 597. Much of the original Greek structure remains intact. Augustine incorporates Plato's theme that there are degrees of reality. The empirical world is viewed as a corruption of a higher realm constituted by the Forms, which Augustine identifies as God's ideas.

The Platonic pillars of Augustine's metaphysics passed through the hands of Plotinus—who elaborated Plato's mystical tendencies. Augustine presents Plato as a philosopher who made as much progress as feasible without the guidance of divine revelation.

In addition to the shadow imagery inherited from Plato, Augustine imports the Persian dualities of light and darkness. Augustine had been a follower of the Persian prophet Mani (215–276) for nine years. Mani taught that the universe is a battleground between the forces of light and darkness. God is light. Satan controls darkness (plus smoke, fire, wind, and water). Darkness is a smoky substance, not a mere absence of light. The parity between light and dark accords with Aristotle's characterization of light and dark as equally fundamental. Following Empedocles, Aristotle analyzes the chromatic colors (red, blue, yellow) as mixtures of the achromatic colors (black and white).

Young Augustine preferred Manichaeism to his mother's Christianity. Mani did not ask for faith. Reason sufficed. Mani was down-to-earth. Christians were otherworldly.

For Mani, God is purely contemplative. God inspires men to fight evil. God does not dirty his hands by fighting evil directly. The Manichaeans think it is empirically obvious that God is not all-powerful. If God were all-powerful, then he would end suffering.

Mani's disciples pressed this objection against the Christians. Manicheans also ridiculed the dogma that God created the universe from nothing. What was God doing before he created the world? Local bishops retorted, "Making hell for people who ask questions like that!"

Prior to the Christians, there was a consensus that something cannot come from nothing. Creation transforms something indeterminate into something determinate. When Plato describes the creation of the world in the *Timaeus*, he portrays the demiurge as molding the world from raw matter. Shortcomings of the world could be blamed on the limits of matter.

Eventually Augustine met Faustus, the great Manichaean leader. Augustine appreciated the charisma of the seer. But Augustine was disappointed by Faustus's lack of philosophical refinement. Augustine concluded that the Manichaeans exaggerated the role of reason in their religion.

Augustine briefly reverted to the Academic skepticism of his youth. But finding that thin soup, he partook of Plotinus's heavier fare. This caused a spiritual indigestion that was only relieved by the emetic of trust.

The freshly purged Christian needed to make a home in the metaphysical landscape Augustine spurned in his youth. Evil was now a pressing anomaly rather than an easy refutation of his mother's simple-minded religion. Augustine was all too familiar with the advantages others enjoyed in explaining evil. In the *Timaeus,* Plato's demiurge makes everything from preexisting matter. Since matter is imperfect, whatever is built from matter inherits its limitations.

But according to Christianity, God created *everything*. Everything God makes is good. Therefore, all matter must be good.

Indeed, everything God makes must be perfect. Yet God cannot make another perfect god because only one being can be all-powerful. So everything other than God must be imperfect.

But this is not to say that these things are bad in a substantial sense. Evil is the complement of good rather than its contrary. There are some things between hot and cold. But there is nothing between good and evil. The mere absence of evil is good. And since what is, is good, 'evil' is an empty complement (just as 'nonexistent' is the complement of 'existent').

Since Augustine denies anything in the middle of good and evil, the absence of the absence of good is good. Thus a recital is commended by exclaiming "Not bad!"

We mark the top end of a scale with 'real': a real wife, a real handshake, a real knife. Someone who is less than a real wife is deficient. Her behavior is measured against the standard set by a real wife.

When you wear a hole through a sock, you only change the shape of the sock. You do not add to the inventory of the world. You cannot make just a hole. You foresee that wearing the sock will create the hole as a side effect. But you do not intend the hole. Similarly, God foresees the privative side-effects of his actions. But this does not make God an evildoer.

Evil lacks *independent* reality. A bone can exist without a fracture but the fracture cannot exist without the bone.

> What, after all, is anything we call evil except the privation of good? In animal bodies, for instance, sickness and wounds are nothing but the privation of health. When a cure is effected, the evils which were present (i.e., the sickness and the wounds) do not retreat and go elsewhere. Rather, they simply do not exist anymore. For such evil is not a substance; the wound or the disease is a defect of the bodily substance which, as a substance, is good. Evil, then, is an accident, i.e., a privation of that good which is called health. Thus, whatever defects there are in a soul are privations of a natural good. When a cure takes place, they are not transferred elsewhere but, since they are no longer present in the state of health, they no longer exist at all. (Augustine 2010b, 3.8)

The privation solution addresses the problem of evil in a way that preserves the sovereignty of God. Instead of there being two opposed forces, as Mani preached, there is only one force. Evil is an illusion that depends on the good. Evil is a hole in good (fig. 13.1). "Woe to those who call evil good and good evil, who put darkness for light and light for darkness" (Isaiah 5:20).

The solution also allows Augustine to preserve a major motif from Manichaeism: the identification of divinity with light. Light is pure. When light contacts feces, the light is not befouled.

Figure 13.1 Evil as a hole in good

In his divine simplicity, God stands to matter as white light stands to the colors of the rainbow. As light passes through drop after drop, light dims. With enough water, as in the Deep Sea, there is an absence of light. Darkness increases with distance from the light. As you sink further from the source of illumination, your shadow grows larger and more distorted.

"God is light, and in him is no darkness at all" (1 John 1:5). Augustine denies that such passages should be read metaphorically: "One does not say that Christ is light in the same sense that he is called a rock, for the first statement is literal, while the second is only a figure of speech" (1982, 136).

Augustine's literal identification of Christ and light is less deflating if the light coming through your window is seen as having a lesser degree of reality than the true light of the divinity. This distinction between divine light and ordinary light was popular in the Middle Ages. Dante's (1265–1290) poetry celebrates the contrast almost a thousand years after Augustine. In book XXVIII of *Paradiso* the narrator is standing at the edge of the Empyrean sphere. He reports,

> I saw a point that shone with light so keen,
> the eye that sees it cannot bear is blazing;
> the star that is for us the smallest one
> would seem a moon if placed beside this point. (1982, 47)

This stronger light revitalizes Plato's myth of the sun. In addition to being the origin of everything, the sun is an ongoing cause of each object.

How reliant on the sun are terrestrial objects for their continued existence? The contemporary answer puts them on a longer leash: if the sun were to disappear, then there would be vast destruction. But not immediately. This is because objects can continue to exist on their own. Physical objects do not need to be nursed along from moment to moment. The later stages of an object are caused by their earlier stages. The object persists because of immanent causality.

Thomas Aquinas (1225–1274) illustrates this existential inertia with a distinction between heat and light. When an object is heated, the object continues to be hot after the source of the heat is removed. The object's continued hotness is due to its past hotness. Light differs. We explain the continued illumination of an object as the effect of an external light source. If the light source were removed, then the object would stop being illuminated.

Now consider something that is completely dependent on illumination—a shadow. One stage of the shadow does not cause the next stage. Shadows have no immanent causality. According to Augustine, the appearance of persisting is an illusion. Despite the appearance of succession, there is only a succession of appearances.

What is the relationship between God and physical objects? According to Augustine, rocks and tables are like shadows. Stages of material things are momentary entities that owe their existence to God: "If He were to withdraw, so to speak, from all things His creative power they would straightway relapse into nothingness in which they were when first they created" (1872, 60).

Most of Augustine's medieval successors did not push the analogy between objects and shadows to this extreme. The momentary view proved more popular among Islamic metaphysicians such as the Persian philosopher Abû Hâmid Muhammad ibn Muhammad al-Ghazâlî (ca. 1056–1111). They appreciated the explanatory merits of the hypothesis that each object is a collection of momentary entities. For instance, it solves the problem of change: how can an object that has one property be identical to a later object that lacks the property? The "perdurantist" solution denies that objects endure through time. Instead objects perdure; one temporal part has the property that a subsequent temporal part lacks. Perdurantists can also solve Zeno's arrow paradox: how can an arrow move given that it is stationary at each moment of its flight? The perdurantist solution is that the motion is just change in place. At each moment, stages of the arrow have slightly advanced positions. The arrow is just a sequence of momentary arrows: → = < ⇢, ⇢,⇢, ⇢,⇢, ⇢,⇢, ⇢ >.

God and His Shadow

Given Augustine's privational account of evil, nothing can be totally evil. A sock can have heel-holes and toe-holes—but the sock cannot be all holes. Augustine boldly concludes that Satan cannot be totally evil. Indeed, Satan is good insofar as he exists.

Augustine is careful to delineate Satan's good qualities. Satan belongs to the highest species—the spirits. And he was the finest specimen of that lofty species. Ironically, this superlative goodness was the cause of Satan's downfall. For Satan's recognition of his merit gave him pride. From pride came envy of God. This led Satan to foment a calamitous mutiny among some of his fellow angels. By rebelling, these spirits became demons.

Other theologians classified demons as a species distinct from angels. But Augustine objected that an evil species is impossible. God would have no reason to create anything purely evil. Instead, demons are spirits that have chosen a lesser good (themselves rather than God).

In John Milton's (1608–1674) *Paradise Lost*, Satan declares, "Evil, be thou my good." But the evil is parasitic on good. What God wills is counterwilled by Satan. Satan is like a defiant but inarticulate toddler. The counter-suggestible boy is keen to say "no." But he must wait for his father to articulate a command. The willful child has a drive to define himself against his father. But the resulting portrait is just a photographic negative of his father.

Despite this derivativeness, Satan's rebellion created an imbalance in the universe. God compensated by creating Adam—and, as an afterthought, Eve. Sadly, Satan succeeded in tempting Eve to disobey the only rule God had imposed. She ate from the tree of knowledge. Eve compounded her sin by persuading Adam to join her.

This original sin put human beings under the jurisdiction of Satan. The Prince of Darkness was entitled to pick souls from humanity as one might pick apples from a tree. God set a trap for Satan by putting Jesus on earth. Jesus, who identifies himself as "the light of the world" (John 8:12), was sinless and so outside Satan's jurisdiction. But God could foresee that the prospect of corrupting Jesus would be so tempting that Satan would overreach. When Satan made the mistake of taking Jesus to hell, Jesus was able to redeem humanity.

Augustine accepted a suggestion by Origen that Satan is Lucifer. According to this interpretation, Satan first appears under a Latin name: "How art thou fallen from heaven, O Lucifer, son of the morning!" (Isaiah 14:12). This actually appears to be an account of a fallen Babylonian king whose Hebrew name is "Helal, ben Shahar," or "Day star, son of the Dawn." The day star is what we call the morning star (actually the planet Venus as seen just before sunrise). *Lucem ferre*, or "bringer of light," became the name "Lucifer." Through the vagaries of biblical exegesis, the fallen king becomes a fallen angel.

As Augustine notes, Christ refers to himself as the "morning star" (Revelation 22:16). Augustine says this metaphor is apt because human knowledge springs from divine illumination. Just as vision depends on light, knowledge of the world and religious truths depends on God.

Satan (as Antichrist) is a shadow of God. He is incapable of God's originality and so is forever copying the divine. The rituals of devil worship are parasitic on those of genuine Christianity. For instance, the Black Mass is a travesty of the Blessed Mass. Augustine explains parallels between paganism and true religion as satanic counterfeiting. Augustine warns against the "noonday demons" that offer a dark light that can be mistaken as genuine illumination. Heretics are persuasive because they take on the appearance of being knowledgeable.

In Augustine's "Lectures on the Gospel of John," Augustine squints to distinguish Christ from Lucifer:

> But Abraham also was born in the midst of the human race: there were many before him, many after him. Listen to the voice of the Father to the Son: "Before Lucifer I have begotten Thee." He who was begotten before Lucifer Himself illuminates all. A certain one was named Lucifer, who fell; for he was an angel and became a devil; and concerning him the Scripture said, "Lucifer, who did arise in the morning, fell." And why was he Lucifer? Because, being enlightened, he gave forth light. But for what reason did he become dark? Because he abodes not in the truth! Therefore He was before Lucifer, before every one that is enlightened; since before every one that is enlightened, of necessity He must be by whom all are enlightened who can be enlightened. (21)

Satan's popularity derives from our desire to personify evil. We want to know what, or better yet, *whom* to hate. Putting a human face on evil gives us the feeling that we know what we are up against. Augustine's privational theory of evil, however, blurs our focus. If everything that exists is good, then we should "love the sinner but hate the sin."

This level of abstraction is hard to sustain emotionally. Given Augustine's privational theory of evil, it is also difficult to sustain intellectually. In hating the sin (rather than sinner) what exactly would we be hating? Given that sin is an absence, there is nothing to hate. Pupils of Saint Anselm's asked why it matters whether they sin. If evil is nothing, then sin is nothing. If sin is nothing, then God will not punish us for sin.

If evil is the privation of good, then how can it harm us? Augustine's answer is reminiscent of the Socratic doctrine that evil cannot harm a good man. According to Augustine, it is the defective *good* that has the causal power. Satan and the fallen angels are self-corrupted creatures who cause grief through their positive features, not their negative features. When a man falls, the bruise comes from contact with the earth, not the absence of support. Or consider the damage done by a man's cut artery. His healthy heart pumps blood through the hole, causing him to bleed to death. The hole does not kill him. His heart does.

Ugliness is a privation of beauty. Reflecting the Platonic identification of truth, beauty, and goodness, Augustine regards each thing as being beautiful to the extent that it exists. As low as a worm might be in the great chain of being, it has some degree of reality and so has some degree of beauty. Accordingly, Augustine is careful to praise the beauty of the worm.

If Augustine's aesthetics is correct, then there are no beautiful privations. So what are we to say about the beautiful shadows cast by the hanging oil lamps in Christian churches? Shadows of crosses radiate from the center, suffusing evening services with delicate, evanescent patterns.

Augustine characterized this kind of beauty as derivative: "For as the beauty of a picture is increased by well-managed shadows, so, to the eye that has skill to discern it, the universe is beautified even by sinners, though, considered by themselves, their deformity is a sad blemish" (1953, 265).

Shadows assist objects of beauty by revealing depth ❑ and texture. But shadows can also step out of the background and be intrinsically beautiful. The shadows cast by decorative lamps manifest beautiful shapes—against a background of light. Perhaps Augustine would insist the light is what is beautiful and the shadows are just a means to the end. Alternately, he might attribute the beauty to the positive properties of the shadow (shape, size, and orientation) rather than to the shadow itself.

God does not create evil. Indeed, evil is, in a sense, unreal. True, there are holes in being. But that is a side effect of making a complete universe, a universe in which every kind of thing gets represented. At the top of the great chain of being is God. Then come angels, human beings, animals, plants, inanimate objects, unformed matter, and shadows. "You do not have a perfect universe except where the presence of greater things results in the presences of lesser ones, which are needed for comparison" (Augustine 2010b, 3.9).

The privational theory of evil was amended in the Middle Ages. Instead of viewing good and evil as contradictories, they were regarded as contraries. Some things are neither good nor evil.

Aquinas consolidated this neutral zone. Although evil is the lack of the good, not every lack is an evil. A man lacks an ability to digest wood. This absence is not evil because it is not a privation (though inability to digest wood is a privation in a termite). Blindness in a man is evil because he is naturally sighted. A fetus develops towards the form of man, and that form is of a visual animal. This connection between goodness and what is natural makes evil something that is always an anomaly, always something in need of an explanation. We ask why a blind man cannot see but not why a sighted man can see. Nor is there any anomaly in a stone that cannot see. Men are meant to see, rocks are meant not to see.

A privation is an absence of something that *ought* to be there. In ordinary usage we let the 'ought' be governed by mere expectation. Aquinas is a natural law theorist. Aquinas thinks the universe is run by natural laws that have both descriptive and normative force.

Aquinas also distinguishes between directed absences and neutral ones. A lump of bronze could become a bust of Plato. Yet the lump's failure to have the shape of Plato's head is not evil. As a lump, the bronze is neither meant to have this shape nor meant not to.

Even with Aquinas's declarations of neutrality, there are recalcitrant examples of evil that appear positive rather than negative. Malice contrasts with callousness in being more than a mere absence of compassion. Pain contrasts with numbness in being more than an absence of feeling. Similarly, some diseases are due to presences rather than absences. The presence of a tail is a disorder for human beings. A tail is a wrong kind of presence for a human being.

The problem with other positive evils is quantitative. Human beings ought to have fingers. But having six fingers is a disorder—even if each is functional by itself.

Perhaps the tail and finger could be characterized as a privation of the correct number of parts (five for fingers, zero for tails). In a similar spirit, malice would be a privation of the correct amount of sadism.

This shoehorning of positive evils into the privation model runs the risk of rendering "Evil is the absence of good" too flexible to be tested by common sense. If we allow ourselves this degree of linguistic license, then we would have no reason to prefer the privational theory of evil over the privational theory of goodness. As we shall see in the chapter discussing Arthur Schopenhauer's pessimism (and perhaps have seen in the chapter on Buddha), the privational theory of goodness has been seriously espoused.

Does 'Nothing' Mean Nothing?

Augustine was puzzled by the Genesis story of creation because it violates the principle that nothing comes from nothing. To minimize its significance, Professor Augustine treats paradoxes about 'nothing' as linguistic confusions.

Augustine concedes that 'nothing' is a peculiar word. Augustine can explain the meaning of 'Adeodatus' by pointing to the bearer of that name: 'Adeodatus' refers to Augustine's fifteen-year-old son. (How did Saint Augustine wind up with a son? In the *Confessions*, Augustine writes: "As a youth I prayed, 'Give me chastity and continence, but not yet.' ") But some words lack a clear referent. Mani thought 'evil' names a substance. But Augustine says 'evil' names an *absence* of goodness just as 'cold' names the absence of heat. Saying evil exists is as misleading as saying cold exists.

Father and son discuss the general nature of words in Augustine's dialogue "On the Teacher." Augustine opens the dialogue with the thesis that all speaking is teaching. Consider a student who asks a question. He is informing his teacher about his desire to learn a particular fact.

Adeodatus proposes singing as a counterexample. His father retreats to the claim that all speaking is *either* teaching or reminding. When Adeodatus insists that some singing is just done for pleasure, Augustine counters that this sort of singing is no different from birdsong and so is not speaking.

Adeodatus acquiesces to this rejoinder. He turns to another apparent counterexample. When praying to God, the Christian addresses an audience he believes to be all-knowing. His father replies that any speaking in prayer is for the edification of onlookers.

Given that all speaking is for teaching or reminding, all words must be signs. Each word should work like a name. Teaching the meaning of a name is a matter of pointing to the bearers of those names. A boy's laughter at "Jesus sat on his ass" (Matthew 21:1–11) shows he needs instruction on the meaning of 'ass'. His mother teaches 'ass' by gesturing at a donkey (*Equus africanus asinus*).

Augustine and Adeodatus immediately begin to puzzle over what each word signifies in a line from Virgil's *Aeneid* 2.659, *Si nihil ex tanta superis placet urbe relingqui* (ii, 3, p. 6) ("If it pleases the gods that nothing be left of so great a city"). *Si* means *if*. We cannot point at *ifs* as we point at asses. When pressed, Adeodatus suggests that *if* signifies doubt in the speaker's mind. Augustine tentatively accepts this suggestion. He urges his son to go on to the second word *nihil* ("nothing"). Adeodatus is forced to say *nihil* refers to what is not. But his father objects: to refer to what is not is to fail to refer. Consequently, 'nothing' should be meaningless. Adeodatus is stumped.

On the one hand, if 'nothing' refers to something, then we would have difficulty refuting this syllogism:

> Nothing is worse than the Devil.
> Nothing is greater than God.
> Therefore, the Devil is greater than God.

On the other hand, if 'nothing' fails to refer, then there is no syllogism. What appear to be premises are meaningless utterances. This denial of meaning fails to do justice to appearance of equivocation. Equivocation requires *two* meanings.

As former teacher of rhetoric, Augustine was aware of how slight pauses in speech dramatically affect meaning. Consider the written sentence: 'A woman without her man is nothing'. Female dependence on men is suggested by the phrasing 'A woman, without her man, is nothing.' The reverse dependence is suggested by 'A woman: without her, man is nothing.' Since the meaning of 'nothing' is dramatically affected by little silences, there must be some meaning of 'nothing' to affect.

Augustine does not use this linguistic evidence to reconsider the presupposition that 'nothing' must have a bearer to be meaningful.

Instead, he urges his son to proceed to the third word of the puzzle, "lest nothing should detain us." More trouble: the third word *ex* ("from") is a preposition. From what does 'from' originate? The name model of meaning continues to force father and son into increasingly quixotic quests for referents. Despite counterexample after counterexample, the pair never question the theory that the meaning of a word is its bearer.

Augustine draws on early theorizing about grammar. For instance, he liberalizes the quarry with Plato's distinction between nouns and verbs. Instead of confining the hunt to objects, he expands the search to include qualities of those objects. Augustine asserts that each sentence has a noun-verb structure. The noun picks out a referent and the verb ascribes some property to that substance.

This noun-verb model became an undebated truism of the Western grammatical tradition. The Scottish philosopher Thomas Reid (1710–1796) listed 'Every complete sentence must have a verb' as among the necessary truths of common sense. Reid does not discuss anomalous sentences such as the one Augustine and his son puzzle over for the name model of meaning. In the case of the noun-verb model, no one debates whether the equational sentence 'This sentence no verb' is grammatical. No one tries to show that the exclamatory existential 'Fire in the hole!' is elliptical for a sentence with a verb.

Doubts about the verb requirement had to be imported from Arab grammarians. They were familiar with languages in which verbless sentences are common. Once these imported doubts are taken seriously, philologists speculated that the substance-attribute metaphysics of the West derives from a grammatical accident of Indo-European languages. Restless philologists pictured one's native language as a prison. They challenged necessary truths by examining the history of languages in light of cross-linguistic evidence.

Augustine's constant anxiety in the dialogue is that a word will be left orphaned without a referent. He searches for something that will always be present when the word is around. One candidate is other words.

'Noun' is sure to refer. After all, 'noun' is itself a noun. Every word would refer to itself if each of its usages was also a quotation of it. Arguably, each use of a word purports to be a model of how the word ought to be used. When asked his name, Augustine would answer

'Augustine'. He simultaneously referred to himself and thereby taught others how to use the name 'Augustine'. Since each word teaches how it is to be used, each word guarantees its own reference.

This insurance scheme must have been especially appealing before the invention of quotation marks. Eventually Augustine becomes sensitive to the distinction between using words and merely mentioning them. If 'bad breath' comes out of Augustine's mouth, it does not follow that bad breath came out of his mouth. A speaker who is unsure of pronunciation can plow ahead without any commitment to being a model of proper usage. When translating, one is not thereby changing the topic.

Translation is faithful only if whatever the speaker had in mind matches what the hearers have in mind. There can be a match only if there are things that match. Thus Augustine begins to regard *ideas* as the referents of words (not the words themselves).

Contemporary philosophers underestimate the originality of this idea. Prior to Augustine, prefacing a report with 'It appears to me' was thought to yield a statement that is neither true nor false. For instance, Sextus Empiricus regards such hedged remarks as retreats into neutrality; no position is being taken. Augustine innovates by characterizing appearance statements as descriptions of the inner workings of the mind.

Appearances provide a sanctuary from the skeptic. They cannot be attacked on the grounds that our senses may be malfunctioning.

Sextus Empiricus was aware of the rhetorical self-defeat in asserting 'I do not exist'. But Augustine turns this pragmatic paradox into a substantive philosophical maneuver. In particular, Augustine argues that skeptical doubts actually support the existence of the self. For given the truth of there being a doubt, there must exist a doubter.

Here we witness an enduring influence of religion on philosophy. Christianity developed in the Roman Empire as a secret religion. The Jews had secured legal permission to worship their God. But as Christianity evolved away from its host religion, Christians were condemned as parasites on the body politic. As pseudo-Jews, Christians lost legal protection. Their strange rituals, such as cannibalistic consumption of Christ's flesh and blood, had to be conducted in the shadows. Given religious persecution, the Christians stressed inward assent over public ritual. Introspection was further encouraged by Jesus's prophecy of imminent world destruction. His followers shifted concern from the accumulation of material goods and status. They concentrated on spiritual self-improvement. Heaven stood as a reward for those with good intentions. Hell loomed as punishment for those with bad intentions. Christians were naturally anxious to monitor the secret state of their souls.

The private nature of ideas seeps into Augustine's conception of speech. Instead of picturing language as a social means of communication, he increasingly thinks of it as a private means of thought. The speaker has unique authority over what his words mean. Consequently, only the speaker is in a position to detect when a metalinguistic explanation is circular and when it is informative: "For discussing words with words is as entangled as interlocking and rubbing the fingers with fingers, in which case it may scarcely be distinguished, except by the one himself who does it, which fingers itch and which give aid to the itching" (1948, 372).

The great advantage of making ideas the bearers of meaning is that there always seems to be a referent at hand. There still remains the question of whether a *suitable* referent has been found. For instance, doubt seems to be a contrived as the referent for 'if'. There may be more plausibility in the proposal that 'if' refers to the idea of 'if'. But the idea of 'if' itself seems as equally hungry for a referent as the word 'if'. And its meal had better not be the word 'if'!

Matters are worse for the proposal that 'nothing' refers to idea of nothing. The *idea* of nothing exists. But how could nothing exist? As we shall see in the next chapter, the riddles of nothingness continued to flourish in the ruins of European universities.

14
Fridugisus
Synesthesia and Absences

After the fall of Rome, the economic decline of Europe deepened. Otherworldly Christians conceived of themselves as in a dark age between the first coming of the messiah and his second. The more terrestrial Charlemagne (742–814) responded to the decline in learning by recruiting scholars. Alcuin of York (735–804), wishing to avoid Vikings, agreed to become headmaster of the Palace School. He taught elementary mathematics and astronomy. His textbook contains the earliest presentation of shunting riddles. Alcuin posed the riddle of how to ferry a wolf, a goat, and a cabbage across a river (without the wolf eating the goat or the goat eating the cabbage).

As an academic administrator, Alcuin presents his domain as a peaceful sanctuary for learning. A crisis erupted when a murderer took refuge in one of the churches under Alcuin's authority. His monks prevented law enforcers from entering the church. Secular authorities argued that the sinner contaminated the sacred space, thereby licensing their entry. They should be thanked for cleansing the space so that its sacred status would be restored! Alcuin defended his monks by arguing that the sacredness of the space persists even when entered by sinners. As Augustine taught, light cannot be polluted by the dirt it illuminates. Alcuin's robust conception of sacred space became entrenched.

Questions arose. The threshold of a church is neither definitely in the church nor definitely out. Are the outer walls of a cathedral sacred or secular? This twilight zone came to be peopled by an upside-down realm of preaching asses, predatory rabbits, and arse-kissing priests. These carnival figures line the flanks and bases of churches. Lascivious monkeys are found at the pedestals of saintly statues. Lewd gargoyles peer down from cathedral columns.

In addition to protecting monks and murderers, Alcuin recruited other scholars. He pleaded to have some of the books he collected at York sent across the English Channel. I do not know whether the Yorkshiremen donated any books. A London acquaintance defined a Yorkshireman as a Scotsman without the generosity.

Alcuin pleas sweetly: "I say this that you may agree to send some of our boys to get everything we need from there and bring the flowers of Britain back to France that as well as the walled garden in York there may be off-shoots of paradise bearing fruit in Tours." Near the end of his life, Alcuin summed up his career: "In the morning, at the height of my powers, I sowed the seed in Britain, now in the evening when my blood is growing cold I am still sowing in France, hoping both will grow, by the grace of God, giving some the honey of the holy scriptures, making others drunk on the old wine of ancient learning."

One of Alcuin's favorite pupils was Fridugisus—who moved from England to France before 796. After Alcuin died, Charlemagne appointed Fridugisus abbot of St. Martin's in Tours. Later Fridugisus was also given the monasteries of St. Omer and St. Bertin. He was the chancellor of Louis the Pious from 819 to 832.

Fridugisus's only surviving scholarship is a letter to Charlemagne entitled "On the Being of Nothing and Shadows." The letter addresses the question "Is nothing something or not?"

Fridugisus first seeks to demonstrate that nothingness is a substance. He then goes on to contend that shadows and darkness are also substances (thereby endorsing the Manichaean account of darkness). Fridugisus divides his proofs into those that rely solely on human reason and those that reason from premises bestowed by scripture.

Human reason can decide whether something is possible. Self-creation is impossible because the creator must already exist to do any creating. Creating something from nothing also seems impossible because there is a missing relatum for z in 'x creates y from z'.

Fridugisus shares Augustine's tendency to model words as names. Disappointingly, Fridugisus does not show Augustine's recognition that the model has trouble finding bearers for words such as 'if', 'nothing', and 'from'. Fridugisus seems ignorant of the grammar that had been taught to Augustine. Fridugisus just plows ahead. General terms such as 'man' collectively denote adult male human beings.

Since 'nothing' is a meaningful word, there must be some *thing* that it signifies.

The linguistic regress of eighth-century Europe is thrown in relief by progress in eighth-century India. Their theorists of negation analyzed 'Ostriches do not fly' as a metacognitive remark rather than a description. The denier of ostrich flight *warns* us not to believe that ostriches fly. Only positive statements describe reality. Negative statements flag intellectual temptations. This metacognitive perspective can be expanded to statements that seem positive, such as the conditional 'If one ostrich has an eye bigger than its brain, then so do all ostriches'. Instead of mapping reality, the conditional tells us how to revise our map of reality upon learning of an ostrich that has an eye bigger than its brain.

Given Fridugisus's theory of meaning, the Bible becomes an inadvertent inventory. Since Genesis relates how God created the world from nothing, nothing must be the most versatile building material:

> For this is [the authority] that declares that the things first and foremost among creatures are produced out of nothing. Therefore, nothing is a great and distinguished something. It cannot be assessed how great is that from which so many and so distinguished things come, since not one of the things generated from it can be assessed for what it is worth or be defined.
>
> For who has measured the nature of the elements in detail? Who has grasped the being and nature of light, of angelic nature, or of the soul? (Fridugisus 1995, 3)

In Plato's *Timaeus* the demiurge creates the world from primordial chaos. For Fridugisus, God creates the world from nothingness.

We do speak as if little nothings can be parts of objects. To turn a bowl into a colander, *add* holes. Shadow is added to a scene to render objects more three dimensional.

The addition of shadows would be impossible if 'shadow' failed to denote anything. Since 'shadow' denotes something, an absence of light, shadows are possible.

Fridugisus's strategy for proving that shadows exist is to show that their existence is implied by authoritative texts. Among medieval Europeans, the Bible enjoyed this canonical status. Accordingly, Fridugisus seeks scripture that implies the existence of shadows.

The Bible opens with many references to shadows. Genesis 1:2 reports "And the shadows were over the face of the deep."

For Fridugisus the biblical passages about shadows do more than establish the existence of shadows. He thinks they also imply much about their *nature*. At this juncture one expects Fridugisus to dwell on miraculous shadows. For instance, laws of nature require a shadow to lengthen in the afternoon. But God intervenes to make afternoon shadows shrink (2 Kings 20:10). The sick are healed by the shadow of Peter (Acts 5:15).

Fridugisus is conservative. All of his descriptions of shadows are intended to corroborate common-sense generalizations about shadows. There is only one mention of a miraculous shadow:

> When the Lord punished Egypt with terrible plagues because of [its] oppressing the people of Israel, he enveloped it with shadows so thick they could be felt. Not only did they deprive men's sight of [its] objects, but also because of their density, they could even be touched by the hands. Now whatever can be touched and felt must be. Whatever must be, it is impossible for it not to be. And so it is impossible for shadows not to be because it is necessary for [them] to be, as is proved from the fact that it can be felt. (1995, 5)

This passage brings to mind a curiosity reported by cave explorers. When plunged in the perfect darkness of a cave, the explorer cannot see his hand in front of his face. Yet when he *waves* his hand, his hand seems to become faintly visible. When his cave-mate waves *her* hand in front of his face, he has no such experience.

Neuroscientists explain this lightless "seeing" as synesthesia (Dieter et al. 2014). Information from the proprioception of limb motion spills into the visual channel. This commingling between the two senses is experienced as a faint hand wave. Strong synesthetes experience a more vivid handwave. The eye-tracking of the hand-seer is smooth, not the jerky motion of subjects merely imagining their hand passing before their face.

Synesthesia is a continuum, with the vast majority of us having no conscious awareness of intermingling of information between sensory channels. However, our understanding of metaphor ("honey of the holy scripture") suggests sensitivity to a synesthetic undercurrent. Synesthesia propels nervous laughter from the joke "Friar Tuck is so fat that his shadow is used to grease the monastery's water wheel."

Fridugisus is drawn to biblical evidence of tangible shadows because touch is an even stronger test of reality than sight. When doubting Thomas sees the resurrected Jesus, he attains full certainty by *touching* the hole in Jesus's body made by a Roman spear.

Normal shadows are not palpable. They do not provide any tactile sensations. Shadows can only be touched in the insentient sense of physical contact (in the way a page in a printed book touches the next page). The toucher of a shadow can be another shadow.

We expect shadows to touch only when their casters touch. But in *The Exhaustive Treatise on Shadows* the Persian scholar Al-Biruni (973–ca. 1052) points out that your shadow-finger touches your shadow-nose *before* your finger touches your nose. The shadow-lips of a chaste couple may kiss without the couple kissing.

The motive for Al-Biruni's focus on shadows was to prevent astronomical fallacies from distorting rituals. Islam's five ritual prayers require daily attention to time and spatial direction. These motives for astronomy and geography were absent from Christianity. The chief stimulus for Christian mathematics was calculating the exact dates for Easter.

In addition to expecting shadows to resemble their casters' actions, we expect them to resemble the casters themselves. This expectation of resemblance overrides what is optically possible (Casati and Cavenagh 2019, chapter 5). Artists pander to this copycat fallacy when they depict a diagonally illuminated cube as having square shadows rather than a hexagonal shadow.

Shadows provide no resistance. They cannot be felt as a breeze can be felt. However, it is not as if shadows retreat from an extended finger. You can contact a shadow. The coolness of a shadow can be felt. As you enter the shade cast by a tree on a hot day you feel relief from the radiant energy of the sun.

Fridugisus infers that shadows are the sorts of things that can be *made* because "in the Lord's passion the evangelist declares that shadows were made from the sixth hour of the day until the ninth hour" (1995, 6). Are all shadows made? If blocking the light is essential to shadows, then it seems each shadow must have a creator.

However, not all darkness is due to light blockage. The biblical creation story assumes universal darkness until God intervenes. Darkness is a default state that can exist purely by the absence of a light source.

The creation story also shows that shadows have origins. They are in time. They must also be in space because the prophet David said, "He sent shadows" (Psalms 105:28). What is sent moves from one place to another. Other passages underscore the locatability of shadows: "He put the shadows as his hiding place" (Psalms 18:11). Jesus himself said, "The children of the kingdom will be cast forth into the shadows outside" (Matthew 8:12).

If Fridugisus had been a common-sense philosopher, he would have appealed to our ordinary practices of tracking the movements of shadows. But these practices are riddled with false predictions. People exaggerate the resemblance of shadows to the objects that cast them. They treat shadows as objects despite having a lifetime of counterexamples. As illustrated by Augustine in his attachment to the name model of meaning, the ability to learn from a counterexample is highly irregular.

The motion of a shadow around the sundial shows that there must be invisible movements of shadows. The gnomon of a sundial is the object casting the shadow which points at the time of the day. To prevent time-tellers from falling on the gnomon, sundial designers transform the time-teller into a gnomon. The time-teller steps on a brick corresponding to the month (fig. 14.1).

He thereby adjusts for the earth's varying tilt toward the sun. Shazam! The human gnomon's shadow rests on a numeral that corresponds to the time of the day.

But the gnomon's shadow does not really rest. Outdoors, the human gnomon's shadow creeps continuously even if he is still relative to the earth. This slow movement might be discernible to a creature with a visual system a different time scale.

Figure 14.1 Human Sundial at the Austin Nature and Science Center

Can there be movements of a shadow that are *universally* indiscernible? Consider the round shadow cast by a disk: ❍. If the disk spins, does its shadow spin as well? If you are looking only at the shadow, you cannot see any difference. The speed of your vision is irrelevant here. If shadows are appearances, then the shadow of the spinning disk does not itself spin.

The more radical way to reach the verdict that shadows do not spin is to deny that shadows persist over time. Recall this punctate thesis—first argued by the ancient Chinese Mohists (a movement that flourished during the Warring States era from 479 to 221 BCE). A flying bird persists because each stage of the bird causes the next stage. The stages of a bird's shadows lack this "immanent causation." Each stage of its shadow is caused by the interposition of the bird and the sun. Thus, each stage of the bird's shadow resembles a page in John Barnes Linnett's kineograph—illustrated here in his 1868 patent (fig. 14.2).

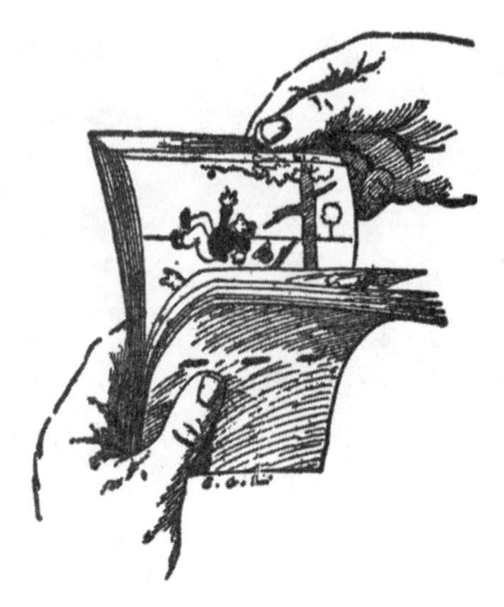

Figure 14.2 The kineograph

One stage of this "thumb cinema" does not cause the next stage. This independence sets the stage for the Mohists' literal interpretation of Chuang Tzu's aphorism "The shadow of a flying bird never moves."

Augustine is sympathetic to the punctate thesis. He intimates that each stage of the universe depends on God's will—making the whole universe akin to a shadow. Islamic theologians welcome the implication that everything depends on God's will at each moment, not just at creation. Their refrain, "God willing," is spoken literally—as a perpetually relevant metaphysical insight. For the Mohammedan, there is no existential inertia. The existence of an object does not continue by default.

Fridugisus refutes Zeno's skepticism about motion by appealing to the authority of scripture rather than by observation of motion, such as turning a page of scripture. There are many verses in which shadows are described as moving. Fridugisus counts these authoritative commitments as better evidence than the testimony of the senses. Pliny the Elder (AD 23/24–79) makes no mention of the three hours of darkness that covered the whole earth during Christ's crucifixion.

Instead of doubting the Gospels, Fridugisus would conclude that Pliny was incomplete in his record of astronomical prodigies.

Rather than directly appealing to observation, Fridugisus appeals to scriptural reports of observations. Recall the Greek debate over who owns the shadow of an ass. Phrases such as "like his shadows" (Psalms 139:12) suggest that shadows can be owned: "If the light that is in you is shadows, how great those shadows will be!" (Matthew 6:23). Since 'shadow' is being used in the plural in this sentence, shadows must be countable. And what can be counted must exist.

What about night? Fridugisus thinks night is made of the same stuff as shadow. Scripture demonstrates that night is a substance:

> When the Lord made the division into light and shadows, he called the light "day" and the shadows "night" (Genesis 1:5). For if the name 'day' signifies something, the name 'night' cannot help but signify something. Now 'day' signifies the light, and light is a great something. For the day both is and is something great. What then? Do shadows signify nothing when the name 'night' is imposed on them by the same maker who imposed the title 'day' on the light? (1995, 5)

Appeals to ordinary language are questionable insofar as they rest on the judgment of human speakers. People introduce words for things that do not exist. Fridugisus cautiously fortifies his arguments with linguistic theodicy:

> The creator stamped names on the things he made, so that each thing would be known when it is called by its name. Neither did he form any thing without its [corresponding] word, nor did he establish any word unless that for which it was established existed. If it were the case [that God had established a word with no corresponding thing, the word] would seem entirely superfluous. And it is wicked to say God has done that. But if it is wicked to say God has established something superfluous, [then] the name God imposed on the shadows cannot appear in any way to be superfluous. (1995, 5)

The word of God cannot be loose talk. From 'nightfall' we can infer night is weighty. And what has weight is substantial.

Fridugisus slavishly relies on scripture. His overarching methodology, however, is the same as that of the revered twentieth-century atheist W. V. Quine. To develop a theory of a given phenomenon the philosopher consults the most authoritative texts. What else can he do if he lacks special expertise? (Recall Nagarjuna's professionalism in restricting himself to connecting the dots, not making the dots.) The philosopher then applies logic to ascertain what the authors of those canonical texts are committed to believing. These commitments represent the best-supported view of the phenomenon.

The deferential philosopher's reliance on canonical texts requires him to distinguish authorities from pseudoauthorities. Genuine authorities need to be weighted by their degree of expertise. As the twentieth century approached, the physicist Lord Kelvin disagreed with the biologist Charles Darwin and the geologist Charles Lyell on the age of the earth. Which field has precedence for the philosopher who wishes to defer? The oracles issue conflicting timelines.

Physics and parapsychology conflict because the subject matter of parapsychology is paranormal events—events precluded by physics. W. V. Quine responded by excluding parapsychology as pseudoscience just as the church fathers excluded many stories about Jesus as apocrypha.

Denominations disagree about which stories get into "the" Bible. This is an ongoing process because the evidence is shifty. There are also discoveries of new *agrapha* (sayings of Jesus that do not appear in the canonical Gospels). In 1950, a respected classicist, Paul Coleman-Norton, published "An Amusing 'Agraphon'" in the *Catholic Biblical Quarterly*. He reports that during World War II, as a military censor in French Morocco, he discovered a slip of paper in an ancient Islamic text. This was a commentary on Jesus's description of hell in Matthew 24:51. Jesus warns "in that place there will be weeping and gnashing of teeth." Intriguingly, the Moroccan fragment continues the description of hell beyond what was already in the Bible. For one of Jesus's listeners asks the question which has occurred to generations of subsequent readers: "Rabbi, how can these things be if they be toothless?" Jesus's answers: "O thou of little faith, trouble not thyself; if haply they will be lacking any, teeth will be provided."

Jesus's drollery was welcomed by journalists. However, another classicist reported he had heard the same *agraphon* told as a scholarly joke prior to World War II. This was discomforting to those who assumed that fraudsters would be deterred by threats of divine punishment.

Scientists stress that fraud will be detected by the requirement that results be replicated. In practice, they mostly rely on trust. One sign is the outrage they express when fraud is exposed.

Even so, Quine believes science is far more credible than any other source. Quine began his career as a nominalist, like Buddha, believing each thing has a position in space and time. Sets are abstract objects because they do not exist at a particular time and place. So Quine initially rejected sets (and numbers, universals, and all other abstracta). After enlightenment by physics textbooks, Quine concluded that some physical statements make indispensable reference to sets. He accepted sets and abandoned nominalism.

Modes of Being and Nonbeing

The Irish monk John Scotus Eriugena (ca. 800–ca. 877) rejected Fridugisus's presupposition that there is only one sense of 'exists'. There are multiple modes of being and so multiple modes of nonbeing.

Nature is the totality of all there is *and* all there is not. Complementing the yang of Augustine's divine light, there is the yin of divine darkness. God emanated from this dark state of nonbeing by negating nothingness. God is a nothing that nothings the nothing. Beings come from nonbeing. These finite double negatives cycle from nonbeing to being to nonbeing. God is the "Beginning, Middle, and End of all things."

But their dynamic Originator stands above finite being and nonbeing. God is an infinite being who stands outside time. He has no *fore*knowledge to threaten human freedom. God is transcendent. This makes God unknowable—even to himself.

15
Maimonides
The Divination of Absence

According to Plato, the Form of the Good has the highest degree of reality. The more real something is, the more knowable it is. Therefore, the Form of the Good is supremely knowable.

Aristotle speaks similarly. We all know what it is like to think. Since thinking aims at knowledge and knowing is best, the best being just thinks. The entity to which all other entities are directed is the quintessential knower. The Prime Mover is all-knowing in the sense that his self-knowledge gives him implicit knowledge of everything. Whereas an abacus implicitly reflects all arithmetic truths, the Prime Mover reflects all truth whatsoever.

Theologians followed the lead of Plato and his famous student Aristotle (who was mistakenly regarded as merely developing his teacher's views). Theology is the study of God—the most knowable of things. Therefore, theology is at the apex of the pyramid of inquiry. Other fields build up to theology. They study the divine design of creation. But theologians study the Creator himself.

The back of the United States one-dollar bill has an image, the Great Seal, that the Masons used to convey to exalted state of theology (see fig. 15.1).

This is the all-seeing eye atop an Egyptian pyramid. The four lines of the pyramid represent the different perspectives from every direction: north, east, south, west. These correspond to the various fields of study. Disciplines such as astronomy, physics, biology, and geology differ dramatically at the bottom. But as these specialists look up, they converge to a single point, at which the Eye of God opens.

Moses Maimonides (1138–1204) inverts the theological pyramid of knowledge from △ to ▽. God is the *least* knowable. The Eye shuts.

Figure 15.1 The Great Seal

Divine Independence

Recall Plato's cave. The shadow of the horse-figurine depends on the figurine. The figurine depends on the horse of which it is a copy. The horse depends on horsehood. And horsehood depends on higher universals. At the apex of the copycats is the Form of the Good. Everything depends on the Form of the Good, but it does not depend on anything. This asymmetry makes the Form of the Good unlike anything else. Whereas other things can have properties by instantiating universals, the highest universal cannot instantiate a universal. Only the Form of the Good has an absolute absence of properties.

Following standard theological practice, Maimonides identifies God with the Form of the Good. But Maimonides complains that Plato misinterpreted his own metaphor. True, the sun illuminates all of creation. But the intense light that illuminates every surface makes the surface of the sun invisible.

To conceive a substance, you must add properties to the blank canvas of imagination. When you attempt a portrait of God, you find yourself "painting" with a dry brush. You can only go through the motions of painting. The lack of a resulting image demonstrates the futility of the project. After all, to assign a property to the highest universal, you would subordinate the highest universal to a universal that was higher than the highest universal. That makes depictions of God blasphemous.

If possibility requires conceivability, God's absence of properties would entail his impossibility. However, it is possible to exist without properties. For the cosmological proof of God's existence demonstrates that God is somehow actual. To avoid an infinite regress, there must be a cause that is itself uncaused. That first cause is God.

By "first cause" Maimonides does not mean first in time. For the universe might lack a beginning. Maimonides means first in a chain of dependence. God need only be a *sustaining* cause (not an *originating* cause). God is constantly intervening as the sun constantly intervenes, emanating the warmth and light needed to sustain life.

In sum, God's position atop the chain of being ensures that he exists and yet also has an unknowable nature. Whereas the existence-agnostic denies that we can know whether God exists, the nature-agnostic denies that we can know what God's attributes are. Maimonides juxtaposes theism with nature-agnosticism.

The Epicurean Challenge

Just as Buddhism is extinct in twenty-first-century Afghanistan, atheism was extinct in twelfth-century Arabia. Indeed, no atheist could be found in any of the Islamic territories extending to Cordova, Spain, where Maimonides was born. His representative atheist was a fossil from ancient Greece: Epicurus.

Epicurus had an official theology. To avoid the Greek stigma of atheism, Epicurus breezily concedes that the gods exist. He goes on to assure readers that the gods do not care about them. Just as mountain-dwellers are indifferent to ants in the valley, the heaven-dwellers are indifferent to the mountain-dwellers. Why should the gods care about us? And why should we care about the gods? Everybody should tend their own garden. Epicurus would ask Maimonides: if we can know nothing of God beyond the brute fact that he exists, then why not stick to what we can know?

A pious man might answer that ignoring God would insult him. Blind praise is safer than no praise.

Maimonides disapproves of such sycophancy. Praise that assigns properties to God is false praise. Since God is perfect, the "praise" will understate his merit. Realizing this, you cannot praise God with words.

Nor can you praise God with pictures. As Anaxagoras quipped, if oxen and dogs could paint, they would depict God in their own image. Archaeologists have discovered rock graffiti in Israeli Palestine seeking blessings from Mr. and Mrs. Yahweh (each endowed with a penis). Depictions of God as a king are left-handed compliments. The Creator cannot be described with properties appropriate to the created. "For God is in heaven and thou upon the earth; therefore let thy words be few" (Ecclesiastes 5:1).

Learned Silence

Maimonides drives this minimization of words to its logical conclusion. Silence is the highest praise we can give God. For confirmation, Maimonides cites Psalm 65:12: "Silence is praise to Thee."

Epicurus also counsels silence. This is the restrained silence of a man who respects the maxim: assert only what you know. Since theology will never put you in a position to know, Epicurus turns his attention to subjects that make for better conversation.

Maimonides' silence is the awestruck culmination of intensive study:

> Glory then to Him who is such that when the intellects contemplate His essence, their apprehension turns into incapacity; and when they contemplate the proceeding of His action from His will, their knowledge turns into ignorance; and when the tongues aspire to magnify

> Him by means of attributive qualifications, all eloquence turns into weariness and incapacity! (1904, 1:59)

Theological sophistication is needed to understand the sublime awe inspired by recognition of God's unknowability. Few of Rabbi Maimonides's brethren had the training or intellectual gifts to follow the Neoplatonic proof of God's unknowability. Erudite quietude is limited to the learned. For those incapable of understanding the futility of speech, something must be said.

Negative Predication

To have an attribute is to be an instance of a higher universal. God is the highest universal.

> Hence it is clear that He has no positive attribute whatever. The negative attributes are necessary to direct the mind to the truths that we must believe. . . . When we say of this being, that it exists, we mean that its non-existence is impossible; it is living—it is not dead; . . . it is the first—its existence is not due to any cause; it has power, wisdom, and will—it is not feeble or ignorant; He is One—there are not more Gods than one. . . . Every attribute predicated of God denotes either the quality of an action, or, when the attribute is intended to convey some idea of the Divine Being itself—and not of His actions—the negation of the opposite. (Maimonides 1904, 1:58)

This passage *appears* to offer knowledge of God through a process of elimination—as in the game of Twenty Questions. In *Mystical Theology*, Pseudo-Dionysius (ca. 500) characterizes the seeker as starting from denials of remote properties ('God is not drunk') and progressing by denying attributes closer to the truth ('God is not wise'). The denials are ever more liable to misinterpretation and so must be voiced out of earshot of the ignorant. As one ascends the chain of being, one reaches "the superessential darkness," all the way up to the lofty "cloud of knowing."

However, the method of elimination presupposes that *distinct* facts are making the answers true. Maimonides's answer to any question of the form 'Does God have property P?' is no simply in virtue of God lacking any property. It is as if Maimonides just cupped his hands over his ears before hearing the last word of the question.

At this stage, negative theology bears a disturbing resemblance to a long sequence of examination instructions that finishes: *Now, ignore all previous instructions, sign your name on the test, and hand in a blank page.*

Objectively, every divine negation of the form 'God is not F' conveys the same message. The amount of information we get from any negative predication is independent of any predicate we choose. Consequently, there is no *accumulation* of clues.

There is a *subjective* difference between divine negations. Before learning about negative theology, the naive questioner regards some positive predications as more credible than others. They are more inclined to attribute wisdom, power, and masculinity to God than to attribute ignorance, weakness, and femininity. Novices are startled by the negations of the subjectively probable attributions (God is not wise, God is not kind, etc.). The shocking denials empty their minds of false conceptions of God.

After you absorb the generality of the principle that "God has no positive attributes," you no longer find any differences in the negations. This source of enlightenment has been played out.

What have you learned about God? Nothing. But you have stopped committing the fallacy of anthropomorphizing God. You have stopped reading the Bible literally. You no longer engage in blasphemous prayers. In short, you have been cured of spiritual disorders.

Maimonides was a prominent physician who believed that there were diseases of both the soul and the body. Philosophy was useful in understanding the patient as a whole. Health is the absence of disease; therapy is a negative activity, like untangling a leash. No dog can untie a man's knot. No man can untie God's Not.

Identity Statements

Identity statements such as 'God is identical to God' do not attribute properties to God. They can be literally true without a negation.

Identity statements are not as trivial as they look. Consider the issue of whether Jews worship the same God as the Christians and the Moslems. This is significant for those contemplating a forced conversion. If God = Allah, then a Jew could at least continue to worship the same God.

Christians say baby Jesus = God. Are they also willing to say God = baby Jesus? Is that God fouling his diaper? Logically, identity is symmetrical. But psychologically, identity statements are asymmetric. Reversing the identity is often shocking.

Maimonides thinks there are crucial differences between Judaism and all other religions. In "Epistle to Yemen" Maimonides characterized Christianity and Islam as superficial imitations of Judaism. They stand to Judaism as a statue stands to the king it copies. There is a resemblance at the surface. But inside the statue is mere stone. Islam and Christianity are imitation religions constructed out of envy. God chose the Israelites and the gentiles know it. That is why the Muslims and Christians savor Jewish suffering.

Action Statements

Although we cannot know what God *is*, we can know what he *does*. God is responsible for all of creation. Maimonides promoted secular study because nature reveals God's design.

Clocks are clocks by virtue of what they do—tell time. If God is God by virtue of what he does—then the nature of God can be known by knowing his function (say, sustaining the universe).

Functional characterization of God fails for two reasons. First, God is the assigner of functions, not their executor. God is the architect of your teeth. But God does not chew your food.

Second, attributing properties of God's effects to God himself is contradictory—akin to attributing the effects of fire to fire itself. Fire softens wax, hardens clay, blackens sugar, and whitens logs. If fire is what fire does, then fire is both soft and hard, and both black and white.

Although study of nature may provide much knowledge of God's effects, it does not provide knowledge of God. Negative theology treats

God as a black box whose inner principles cannot be divined from its output.

Study of the physical universe is not sufficient for a *religious* attitude. Twenty-first-century physicists believe that everything is the effect of the Big Bang. Their awe does not translate into religious praise of the Big Bang as the First Cause. Indeed, the fact that the Big Bang is cognitively inaccessible to physicists has an Epicurean effect on them. Physicists condemn speculation about what happened before the Big Bang as futile—and perhaps even meaningless. Under the "no boundary" hypothesis of quantum cosmology, time does not exist. By exploiting imaginary numbers, Stephen Hawking and James Hartle, massage time into just another direction of space. In the process, the Big Bang loses its character as a creation theory. After all, if there is no 'before' and 'after' asking 'What happened before the Big Bang' is like asking 'What is north of the North Pole?'. (Hawking and Mlodinow 2010, 133–136)

The False Hope of Verisimilitude

One might claim that 'God is wise' is at least closer to the truth than 'God is ignorant'. If there are degrees of reality, then there are degrees of truth. So perhaps positive theology could be revived in an incremental way.

Pseudo-Dionysius allows "inadequate affirmations" such as 'God is loving—but not in the way of creatures'. He also allows an even higher stage involving superlatives that tacitly negate possession by any created being: 'God is infinitely knowing', 'God is good beyond excellence', and so forth.

Unfortunately, verisimilitude requires a common scale. This commensurability explains why statements that only have a degree of truth can be hedged into statements that are fully true. Given that Ben is 29, 'Ben is 30' is merely close to the truth. But 'Ben is around 30' is fully true.

If God had an age, then it would be closer to the truth to say he is a thousand years old then to say he is a hundred years old. God is portrayed as an old man with a gray beard. But all depictions of God's age are false. God is ageless. God is off the scale of age in years but not because his age equals some number too large to contemplate. That would still leave God on the scale. Consequently, all positive attributions of an age are equally false.

Metaphor and Fiction

Metaphor typically incorporates an analogy vaulting from the more familiar to the less familiar. An individualist who says, 'Israel retaliated' denies commitment to Israel being an agent. The individualist explains that his remark only means to convey the fact that appropriate Israeli officials acted in a way that made Israel resemble a retaliator. Similarly, 'God flooded the earth' is a metaphorical extension. People are the primary agents. God is a metaphorical agent.

In his treatise "On the One True Agent," Al-Farabi (ca. 900) runs the analogy in reverse. The only agent is the one who can bring being from (absolute) nonbeing. Human beings can only bring being from potentiality. So God is literally an agent and people are only metaphorically agents!

In twentieth century, fascists politicized the inversion. *Fascis* is Latin for "bundle." Each wooden rod is weak. A bundle of rods is strong. Agency requires strength. Consequently, the agent is the collective, not the individuals that compose it. Italians come and go. They are made Italian by Italy, a nation that persists over many generations. The moral community is a community of agents. And those agents are nations, not individuals. What matters are nations, not their nationals.

When discussing the Olympian theology of the ancient Greeks, the monotheist says that there are many gods: Apollo, Brizo, Charon, . . . The monotheist is consistent because Apollo is a fictional character. 'Apollo is a god' means that the monotheist *pretends* 'god' applies to the fictional character. When God appears in a fiction, he is an immigrant object like the sun, rather than a fictional character. Those who deny that manhood can be literally predicated of God could still pretend that manhood is ascribed to him.

Interaction Words

Some adjectives interact with what they modify. 'Healthy' can apply to food, exercise, habits, people, and their urine. These items do not need a common property for 'healthy' to apply to each. Similarly, 'good' has no meaning apart from what it modifies. Arguably, 'God is good' and 'Pleasure is good' can both be true without a common attribute and without equivocation.

There is no reconciliation with negative theology if the metaphor reduces to another positive attribution. Under the simile theory of metaphor, 'God is alive' means that God resembles a living organism. The only way to resemble something is to share some property with it. The conjunction of negative theology and the simile theory of metaphor precludes even metaphorical descriptions of God.

Not all figures of speech that mention God are about God. Friedrich Nietzsche's slogan "God is dead" is about *belief* in God.

The rhetoric of physics uses 'God' to personify the laws of nature. When Albert Einstein said, "God is cunning but He is not malicious," he was doing physics rather than theology. Instead of interpreting Einstein's "God does not play dice" as a trivial truth, physicists read in an objection to Niels Bohr's theory: contrary to the random events posited by Bohr's quantum mechanics, the universe is deterministic. When Bohr retorted, "Stop telling God what to do!" Bohr was admonishing Einstein for his rationalist physics, not his impiety.

Noncognitive theories of metaphor avoid any commitment to a literal paraphrase by comparing metaphors to inebriants. There is a mental change but not one wrought by the information conveyed by the metaphor. The point is to jar the audience in a favorable way.

Is Existence a Property?

If existence is a property and God has no properties, then God does not exist. Some negative theologians respond that this is true for one sense of 'exist' and false for another sense.

They are inspired by Aristotle's doctrine that there are many ways to be. If one further believes that a predicate cannot apply across different categories of being, and places God in his own category, then God would exist in a sense distinct from anything else.

If one further holds that misassigning a predicate in the wrong category ('God yawns') yields a meaninglessness utterance, then positive statements about God are nonsensical rather than false.

This jeopardizes Maimonides's concession that negative predications are true. "Negating" a meaningless statement yields another meaningless statement.

A second difficulty with treating 'exist' as ambiguous is that there is no longer any way to rank creatures in a great chain of being. Each sense of 'exist' would be autonomous. Consider those who assert that only physical things exist but then go on to concede there exists an even prime number. To avoid contradiction, they say that numbers exist in a different sense. Given that 'exist' is ambiguous, 'Chimes and primes exist' is a pun. Ordinary people are tempted say that chimes exist in a stronger way than primes. And Plato is tempted to say the reverse. But the ambiguity of 'exists' deprives both sides of a common scale to make either comparison correct.

A more promising way to salvage the truth of negative predications is to maintain that existence is a precondition for having properties rather than a property itself. Under this view, there is no pun in 'God and His creatures exist'.

The Name of God

'God' is used as a predicate in 'Apollo is a god'. Maimonides should therefore conclude God is not a god. "The gods too are fond of a joke," observes Socrates (in Plato's *Cratylus* 406).

'God is God' is not affected because 'God' is being used as a name. It has the logical form of the identity statement $g = g$ instead of the predication Gg. Under this analysis, 'God' is just a label for God that bears no semantic content.

Some theists defend God's existence by making existence easy. For instance, they say that the negative existential 'God does not exist' is self-defeating because one can refer to God only if he exists.

This strategy might suit a polytheist. But monotheists deny the existence of other gods. This negative aspect of monotheism led Roman polytheists to lump together monotheists with atheists as numerically insignificant. These extremists both denied the gods needed to fortify secular authority.

Esotericism

Since books were expensive, boys were required to memorize the Torah before they understood what it meant. Comprehension can be partial. A boy can use 'father' competently before he learns, on his wedding night, what fatherhood entails.

Just as an infant must be nourished by milk before he can be nourished by meat, the student must be educated by parables before nourishment by principles. The student must defer to the few who can attain higher-level knowledge. Sages, in turn, must guard the piety of the ignorant by refraining from statements that are apt to be misinterpreted.

Instruction comes at different levels. Since Maimonides had to advise a much wider audience, his writings need to be interpreted in terms of their intended audience. Maimonides did not want to jeopardize the piety of those who were apt to misinterpret what he said. He had to build bridges to salvation that even the most ignorant could safely cross. The ignorant were not obliged to understand. They were only required to defer to those who knew better.

Maimonides believed Adam knew proofs of God's existence that were later forgotten. Instead of having a cumulative model of knowledge, Maimonides believed that knowledge was frequently lost because of catastrophes, sabotage, and negligence. He did not accept Plato's view that knowledge is virtue. Thanks to original sin, people act immorally even when they know how they ought to act. Worse, some are attracted to wrongdoing *as* wrongdoing. As a boy, Saint Augustine stole pears *because* theft was wrong, not in spite of the wrongness.

Moses Maimonides was not an intellectual communist. He differed from scientists who share all they know. Like the Egyptian priests,

Maimonides was proprietary. There were religious secrets that ought not to be revealed even to other sages.

Interpretive difficulties are further exacerbated by Maimonides's endorsement of prudent dissimulation. During the turbulent era of the Crusades, tides of religious intolerance rose and ebbed in contested regions. Jews often faced a choice between conversion and death. Controversially, Maimonides said it was better to pay lip service than to be martyred. A Jew could wait for a return to tolerance or at least the opportunity to emigrate. (Jews in Christian areas were more likely to advocate martyrdom, partly in emulation of the Christian ideal of martyrdom, and partly because conversion to Christianity was viewed as polytheism.)

Maimonides himself lived the life of crypto-Jew. He took the religious oath making him a Muslim. He prayed at mosques. He studied the Koran. Maimonides chose to lie about sacred matters. He suborned his fellow Jews to do the same.

One might wonder whether we can trust anything Maimonides says! Should such a prevaricator be accepted as a witness in a trial? Should he be eligible to serve as a judge? One answer is conveyed by a portrait over the gallery doors of the United States House of Representatives Chamber (situated among 22 other lawgivers) (fig. 15.2). A translator of Maimonides's *Guide for the Perplexed* answers in the reverse spirit. Leo Strauss (1952) contends that the likelihood that Maimonides held a given thesis is inversely proportional to how well he argued for it. According to Strauss, Maimonides's arguments in favor of creation from nothing are a sign that he covertly believes in the eternity of the universe!

Strauss's doublethink approach has been surprisingly influential. Doublethink makes all interpretations of Maimonides self-stultifying—including the view itself. Scholars inched away from Strauss. Most now presume Maimonides meant what he wrote.

Maimonides says we should be very humble. His views about our lowly abilities to discuss God reflect this extreme modesty. Critics of negative theology are put in mind of competitive humility in synagogues. The shamus is the temple handyman. Above him is the cantor, who leads the congregation in prayers. At the pinnacle is the rabbi, who teaches and adjudicates issues of religious law. Rabbis

Figure 15.2 Portrait of Maimonides at the US House of Representatives Chamber

worry about appearing vain. A particularly earnest rabbi kneels and puts his forehead to the floor and proclaims, "Before you oh Lord, I am nothing." Impressed by the rabbi's piety, the cantor kneels alongside the rabbi, puts his forehead to the floor, and says, "Before you oh Lord, I am *less* than nothing!" The shamus then joins the pair on the floor, and says, "Before you oh Lord, I am nothing." The cantor nudges the rabbi and says, "Look who thinks he is nothing!"

VI
SCIENTIFIC NOTHING

16

Bradwardine

Absence of Determination

Geoffrey Chaucer's "The Nun's Priest's Tale" begins with praise for Augustine, Boethius, and Thomas Bradwardine. As the former archbishop of Canterbury, Bradwardine is the local favorite for the *Canterbury Tales*.

Bradwardine was the shortest-lived of archbishops. The Black Death abbreviated his reign in 1349, just five weeks after consecration by Pope Clement VI, exactly one week after returning to England. Bradwardine's work on free will bore on the great issue of Chaucer's day: does God's freedom preclude others from having freedom?

Freedom as a Zero-Sum Game

For every freedom, there is an equal and opposite limitation. Your right to a slice of the pie is a duty upon others not to eat it. As *your* liberty increases, *their* liberty decreases. When your gains are added to their losses, the sum is necessarily zero.

If God exists, he is all-powerful. Anything all-powerful has total freedom. Since God has all the freedom, no one else is free.

Bradwardine accepts this calculation for *primary* freedom. People have *secondary* freedom to ratify God's actions. Each of us is a wheel of a cart. No wheel has primary freedom of movement. A wheel may have the secondary freedom of being responsive to the primary motion of the horse pulling the cart. Wheels are designed to turn freely. However, they may warp, rub, and jam. When the Wheelwright mends the wheels, the wheels regain their freedom. Secondary freedom is fulfillment of function.

John the Baptist says of Jesus, “He must become greater; I must become less” (John 3:30). Jesus in turn, when relating to God as son to Father, emptied his will, becoming nothing, a slave (Philippians 2:7). This self-effacing doctrine of humility doubles as a doctrine of secondary freedom.

Those who believe determinism precludes free will say that you are responsible only if you could have done differently. Bradwardine disagrees. Suppose Lucifer will induce kleptomania if you fail to steal this book. Conveniently for the Prince of Darkness, you steal this book voluntarily. Lucifer remains idle. Although you could not have done differently, you are responsible for the theft. What counts is your intention.

How God's Freedom Structures Space and Time

In 1913, Lawrence of Arabia solved the riddle, “How do you collar a leopard?” His young leopard, a gift from a government official in Jerablus, had outgrown its collar. Lawrence's leopard deterred trespassers. But it had no loyalty to Lawrence. Enticing the leopard into a box had made the big cat even nastier. Lawrence did not want to put his hand in the box. He stuffed the cage with canvas bags until the leopard lost freedom of movement. Lawrence then opened the top of the box, and put the new collar on the wedged leopard.

Lawrence had exploited the relationship between space and freedom. Bradwardine associates space with God's liberty. God's elbowroom elbows out everyone else's elbowroom.

Indeed, God's omnipotence controls the nature of space itself. Primary freedom requires a total absence of constraint. Any shortage of space would be a constraint. Thus total freedom requires an infinite theater of operation. Contrary to Aristotle, the universe must be infinite.

What goes for space goes for time. God must have been free to create the cosmos earlier or later. Contrary to Aristotle, there must have been a precreation void.

God must be free to have positioned the universe elsewhere. Consequently, Aristotle is mistaken about there being a center of the universe. The Christian scores the debate on the void: Jesus 1, Aristotle 0.

Holes in Space

To be free, God must have the liberty to annihilate any body while also restraining other bodies from occupying the vacancy. Albert of Saxony noted that this raises a question about distance. Suppose everything beneath the sublunary sphere were annihilated. That would leave absolutely nothing (not even air or some subtler gas). Further assume that the sphere does not shatter (perhaps because it is uniformly thick so there is no more reason for the sphere to break

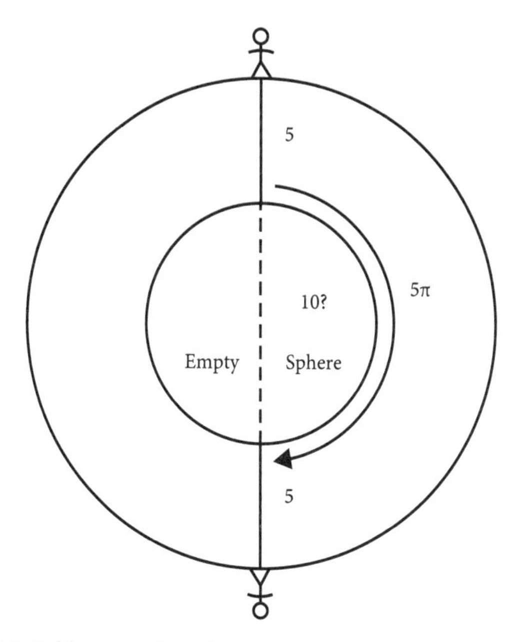

Figure 16.1 Sublunary sphere figure

at one point rather than another). What is the *distance* between the two observers who stand five units above and below the 10-unit sphere? (fig. 16.1). Initially, we answer that the two observers stand 20 units apart: 5 + 10 + 5 = 20 units. But then we remember that the sphere does not contain anything. So there is no substance inside the sphere to which we can apply the quantitative phrase 'has a 10-unit diameter'.

If we conclude that there is no distance within the sphere, then Albert will ask about the two points at opposite poles of the sphere. If there is no distance between them, then they are in contact!

Responding to this threat of collapse, we instead deny that the concept of distance applies *across* the gap. But now surfaces our awkward ability to measure the distance *around* the gap: circumference = $2\pi r$, so half the circumference is 5π, or 15.5 units.

Henry of Ghent argues that the void in the sublunary sphere must have dimensions: the void's dimensions can be aligned with a measuring rod. Further, there is a *shape* to the void. In particular, the void shares the roundness of its spherical container.

A sphere is left-right symmetrical. But if the container has the shape of a left-handed glove, the void will have an orientation that differs from a container that has the shape of a right-handed glove. A hand that snugly fits into one glove will not fit into the other. Henry of Ghent's attribution of a shape to the void can be expanded to an attribution of orientation. Two voids could have same size and shape and yet differ as much as the gloves.

Jean of Ripa (fl. 1357–1368) notes that the spatial anomalies of the void spawn puzzles about motion. Suppose God successively annihilates the cells between A and B (fig. 16.2). If each cell is one unit in length and distance requires a subject to have that quantity, then the number of units between A and B decreases as the cells are annihilated. Yet A and B are at rest!

A	1	2	3	4	5	B

Figure 16.2 Annihilation of intermediate cells

The Logical Limits of Omnipotence

The only limit on God's power is logical. He cannot make a circle with corners. Nor can God, as a perfect being, do anything imperfectly. God cannot testify falsely. Nor can God do anything that implies an indirect contradiction. In particular, God cannot create an actual infinite. For then God could create another God. The possibility of twin gods would imply the possibility of a contest between an irresistible force and an immovable object. Such a contest would imply a contradiction.

Since God's freedom requires an actual infinite space and God could not have created an actual infinite, infinite space must be an attribute of God. Space is the incarnation of God's omnipresence. Thus the nothingness behind free will explodes into the nothingness of infinite space and time. All freedom trails back to a single, uncaused being who is responsible for everything. Human beings are free to the extent that they endorse acts that flow from God's freedom. Those who will what God wills, get exactly what they desire—and ought to desire.

Black Plague Liberation

Free will was a lively issue in the fourteenth century. On the one hand, God's freedom seemed to fate everyone else to a destiny over which they had no control. On the other hand, agricultural disaster had opened unprecedented opportunities and hard choices. Unseasonably cold weather precipitated crop failures. Crop failures caused famine. Famine caused plague. Plague caused pogroms. Pogroms caused war. A common solution to problems is compassion. But the Black Plague selected against compassionate people. Those who nursed the sick sickened. Priests fled. Then spouses, then parents. Hard-hearted survivors eventually inherited property, rank, and an improved market for their labor.

Formerly, the church had figured as an ultimate arbiter of disputes. The plague blunted the pyramid of clerical power. In 1377, Urban IV in Rome and Clement VII in Avignon each claimed to be pope. Christians had to decide who was pope and who was antipope.

A hundred years earlier, the Condemnation of 1277 had banned 217 doctrines associated with Aristotle. The condemnation was organized by followers of Augustine and aimed at the Paris schools of theology. Thomas Bradwardine was educated in this period of Augustinian reform. Augustine taught that we are all tainted by original sin. Through the sacrifice of Jesus, Christians may receive salvation by the grace of God. But non-Christians (Jews, Muslims, pagans, and even unbaptized dead babies) go to hell. Good works themselves cannot cause a completely autonomous God to do anything. God is not a judge whose verdicts are *effects* of how the defendant behaved. The verdict is already decided before the defendant exists.

Augustine wavers. He lodges unbaptized babies in the nicer apartments of hell (next to Socrates, Plato, and other pre-Christian philosophers). Augustine suggests that good works might be a *symptom* that one is among the elect.

Some suggested that good works could be motivated by the good news they generated. If you help your feeble neighbor, then that is evidence you were previously chosen to be among those who will go to heaven. If maximizing good news (independent of causal impact) is rational, then the timing of the reward is irrelevant. The incentive would work regardless of whether heaven comes before or after your terrestrial existence. Suppose you forgot, at birth, whether you had been heaven or hell. You would still gain clues by whether you did good works. Sacrificing to raise the probability of a past event is irrational. Sacrificing to raise the probability of future event is rational only if the sacrifice is likely to be a *cause* of that event.

Causal decision theory (which endorses the option with the best *effects*) assumes an objective difference between past and future. Augustine has a subjective theory of time. The past is what can be *remembered*. The present is what can be *perceived*. The future is what can be *predicted*. Since God perceives everything at once, everything is now to him. Omniscience is not *fore*knowledge for God.

Soft determinists claim 'Every event is caused' is compatible with freedom. Chaucer was familiar with this doctrine through his translation of Boethius's *The Consolations of Philosophy*. Consider two readings of 'If the librarian foresees that you will return a book, then you must return the borrowed book':

Necessity of the Consequence: Necessarily, if the librarian foresees that you will return the book, then you will return the book.

Necessity of the Consequent: If the librarian foresees that you will return the book, then necessarily you will return book.

The first conditional is true but the second is false. The librarian's foreknowledge does not compel you to return the book. After all, the basis of her prediction could be the very fact that you are free. The absence of obstacles and disabilities means that you, as a conscientious patron, will keep your promise to return the books.

A Divine Monopoly on Freedom

Agents presuppose they originate their own actions. You are not like a pen that merely transmits the free motions of the author. Your felt autonomy is the basis for human dignity and emotions such as pride and guilt.

Chaucer tilts the playing field against human freedom by having barnyard animals debate *their* free will. The tale begins with a groan from the sleeping rooster Chanticleer. The favorite of his seven wives, Pertelote, wakes him. Chanticleer recounts a nightmare in which he is attacked outdoors by an orange, hound-like creature resembling the beast that attacked his father and mother. Chanticleer fears that the dream is prophetic. If his dream foretells the future, would it be prudent of him to stay in the chicken coop today?

This precaution seems futile given the tautologous nature of the conditional 'If Chanticleer's dream of the attack is prophetic, then Chanticleer will be attacked'. By definition, all prophetic dreams are dreams that become true.

Chanticleer recalls Boethius's diagnosis that this form of argument mistakes the necessity of the whole conditional, $(p \rightarrow q)$, with the necessity of the consequent which is inside conditional $(p \rightarrow q)$ Chanticleer admires the distinction. However, he is a practical chicken who does not clearly see how Boethius's scope distinction solves the problem.

Bradwardine did understand how Boethius's distinction shows how librarian's foreknowledge of you returning the book is compatible with you freely returning the book. But God's foreknowledge of you returning the book is another matter. What God foreknows, he also forewills. Since God is omnipotent, God differs from a passive observer who foresees a collision between two blind pedestrians. God is actively and intimately involved. God dictates each step. Indeed, God's omnipotence ensures that he takes each step. The pedestrians' steps are like steps of a marionette. The puppet master is walking the pedestrians. The blind pedestrians only differ from being puppets in having a will that can submit to the steps—and the collision.

Any attempt by Chanticleer to avoid his fate by hiding in the chicken coop will be just as futile as an effort to change the past. Chanticleer can no more save himself than he can save his dead parents.

If free will requires absence of causation, then only uncaused beings are free. Like other library patrons, you feel like an uncaused being when you decide to return a borrowed book. You feel that the act originates with you. According Bradwardine, however, you no more originate your action than your shadow originates its movements. You are completely dependent on God.

For Bradwardine, freedom is a fixed quantity. What freedom one agent gains, the other agents lose. If I am free to use the northern half of a pasture, you are thereby restricted to the southern half. The best distribution of freedom would be in accordance with who makes the best choice. Since God is the best decision-maker, he should have all the freedom. Apportioning any freedom to man encroaches on God's liberty.

According to Pelagius we are able to freely follow God's law and thereby merit salvation. Pelagius denied Augustine's account of original sin. Augustine petitioned the pope to declare Pelagius a heretic—unsuccessfully.

Bradwardine's theology is a systematic recantation of his youthful Pelagianism. Bradwardine limits human freedom. As long as God is doing just what you wish, you are as free as a sailor riding a swift current home. You are captain of your fate, master of your destiny. Freedom is conformity.

Many theologians minimize God's participation in sin. Concern for his holiness leads them to say that God is only a *remote* cause of crimes. God causes crime only in the way that honest labor causes crime (by creating products that can be stolen).

Bradwardine spurns this insulation: God is the immediate cause of *every* event, even sinful acts.

The impotence of the sinner fails to exonerate him because it is still true that he would have caused evil if he had the power. God is the farmer who slaughters a hen that would have been pecked to death by a bully chicken. The bully chicken is the real culprit—even though the Farmer does the actual killing.

Despite Chanticleer's sympathies with Bradwardine, Chanticleer acts as if he believes he is free. He is somewhat assuaged by Pertelote, who suggests that he merely needs a laxative. Her exit from the coop also inspires his ardor—which he consummates repeatedly in the yard.

Now relaxed and confident, Chanticleer saunters further afield. He spots an orange animal contemplating a butterfly. Noticing Chanticleer's instinctual wariness, the vulpine lepidopterist explains that the purpose of his visit is to witness Chanticleer's renowned crowing. Chanticleer puffs with pride. As requested, Chanticleer stands on tiptoe with neck outstretched and eyes closed, to concentrate on his cock-a-doodle-doo.

Pounce! Reynard the fox snatches Chanticleer by the throat, slings him over his shoulder, and hauls the terrified rooster into the woods. Percolate's wail alerts the other farm animals. Geese honk. Hogs squeal. Cows moo. Dogs bark. Even the farmer joins the hullabaloo of the fox hunt.

Reynard has a safe lead. Nevertheless, the fox is put on edge by all the verbal abuse. Chanticleer advises the fox: "Sire, if that I were as ye, Yet sholde I seyn, as wys God helpe me, 'Turneth agayn, ye proude cherles alle'!" The rattled fox heeds the voice from his mouth, and upbraids his pursuers. Released from the fox's jaws, Chanticleer flies up to a tree.

The fox apologizes to Chanticleer for the rough handling. He urges his friend to fly down to talk through the misunderstanding. Chanticleer replies by a heartfelt, universal denunciation of vanity. Seeing that Chanticleer has learned his lesson, the fox can only match Chanticleer's regret with a heartfelt denunciation of boastfulness.

Plight of the Void

Chaucer has been compared to Euripides (ca. 480—406 BC) because he sensitively portrays the diversity of ordinary people. Nevertheless, Aristotle criticizes Euripides for frequent recourse to the deux ex machina plot device (*Poetics* 1454a33–1454b9). This was named after the crane that lowers the problem-solving god into the play.

According to Aristotle, a drama should develop naturally from its initial premises. Playwrights ought not to abuse their right to stipulate the course of events. Aristophanes makes the same point in his parody *Thesmophoriazusae*. Whenever difficulties accumulate, Aristophanes lowers Euripides into the play.

Aristotle would have been more chagrinned by a deux ex machina orchestrated by the bishop of Paris in 1277. The thirteenth century began with the void nearly extinct in medieval Europe. Muslims had preserved Aristotle's physics and written erudite commentaries. This scholarship fell into the hands of Christians in their reconquest of Spain. The church subsidized intensive study of the manuscripts to catch up to the infidels.

Nothing competed with Aristotle's physics. And nothing was losing! Aristotle's opposition to the void was overwhelming. With the benefit of his Islamic allies, Aristotle persuaded Christian intellectuals that the universe was a plenum. Everything is filled, so there are no tiny voids. There are no middle-sized voids. Nor can such voids be artificially created.

If misery loves company, then the void could find grim comfort in the plight of Aristotle's other target: infinity. The linked fates of the void and infinity are epitomized in a carpentry thought experiment. Albert of Saxony selects an infinitely long beam of wood with a square cross-section of one inch by one inch. He saws off a one-inch cube. This leaves just as much wood. Albert saws off eight more cubes to pack around the first cube. Now he has a 3 × 3 × 3 cube (fig. 16.3). Gaining momentum, Albert saws yet more cubes to surround this cube, yielding a 5 × 5 × 5 cube. On and on, he saws. Ultimately, Albert forms a giant cube that fills the entire universe. Lesson: infinite objects violate the conservation of matter.

Figure 16.3 Filling space with cube cut from an infinitely long rod

The Actual Infinite Void

If a Christian were forced to choose between infinity and the void, then he might favor the void on the grounds that only it has direct biblical support. Recall Fridugisus! In neither the Old Testament nor the New Testament is God described as infinite. However, the void gets first mention in the Genesis. Recall how Fridugisus, in the eighth century, parlayed these many references to absences into a theory of shadows and nothingness.

The Bible attributes powers to God that presuppose some empty space to maneuver. Genesis begins with God freely choosing to create the world. God could have chosen to create the world in a different spot. He could have made the void bigger or smaller.

Indeed, God could have chosen to make the universe a different shape. And if that shape were asymmetric, then God could have chosen a different orientation. A cosmos in the shape of the letter *b* could have been left-right reversed as *d*.

Al-Ghazâlî managed to see this possibility of orientation shifts despite assuming that universe is *spherical*. Orientation is only relevant to a moving sphere—as when we distinguish between clockwise and anticlockwise motion. This question does arise subtly for al-Ghazâlî because his universe is a system of rotating spheres. Al-Ghazâlî notes that the highest sphere moves east to west. The highest sphere could have instead moved west to east. If the lower spheres moved in the opposite direction, the same configurations of heavenly bodies would occur. But a change of orientation must be relative to something else. So if all bodies could collectively change orientation, then there must be something beyond bodies that provides the reference frame.

Immanuel Kant would later formulate al-Ghazâlî's east-west reversal as an argument for the existence of absolute space: if the universe contained nothing but a hand, it would be either a right hand or a left hand. So space itself must provide the reference frame that makes the lone hand a left hand rather than a right hand, or vice versa. Therefore, space must exist independently of the objects it contains.

Aristotle believed that earth is attracted to a unique point in space. The center of the universe. This unique attraction-point explains why Earth is at the center of the universe. Center-attraction explains why there could not be more than one Earth. If there were multiple Earths, they would collide.

God could have refrained from creation and left the universe empty. He was free to create the world at any time he wished. This implies that there was a universal void prior to creation. Given that time has no beginning, this void must have been infinitely old. Christian commitment to omnipotence requires that the void is not only possible but also actual. The precreation void and the extracosmic void would be actual voids that were also infinite—a double violation of Aristotelian doctrine.

The Condemnation of 1277

Low-level church authorities tried to enforce Christian orthodoxy on the Aristotelian philosophers. Their failure was underscored by Thomas Aquinas's influence at the key theological schools in Paris.

The bishop of Paris, Stephen Tempier, saw an opportunity when Peter of Spain became Pope John XXI. Peter was a logician who favored Plato over Aristotle and therefore Augustine over Aquinas. Bishop Tempier promptly received a commission to investigate the Paris schools. Tempier exceeded his orders by mega-banning 219 articles associated with the Arab Aristotelian philosopher Averroës. The intervention was understood as a veiled condemnation of Aquinas.

Pope John XXI died a scholar's death: a month after election, his specially constructed study collapsed upon him.

His ban survived. The scholars in Paris were forced to make room for the void.

Many begrudged their billeted guest. The scholars confined the void to the realm of mere possibility. God *could* annihilate a body but would not actually do so. There would be no point in interfering with his own creation!

A more concessive line was to distinguish between the supernatural and the natural. Aristotelian principles held by default except in the case of miracles (which some theologians quarantined to an earlier age of miracles).

But a much greater concession was needed. The precreation void and the extracosmic void showed that the void does not depend on bodies for its existence. The only candidate is God himself. Somehow, the most negative thing must be intermingled with the most positive thing.

The Augustinian Embrace of the Void

According to Augustine, God embraces the world

> in every part and penetrating it, but remaining everywhere infinite. It was like a sea, everywhere and in all directions spreading though

> immense space, simply an infinite sea. And it had in it a great sponge, which was finite, however, and this sponge was filled, of course, in every part with the immense sea. (2010a, 168)

Augustine's metaphor may have been inspired by Acts 18:27–28: "For in him we live and move, and in him we exist."

Space is, as it were, the body of God. Every body is embedded in this half-physical, half-spiritual medium. Even if these bodies filled space, God would not be identical to them. Augustine's vision opens the enthralling possibility that God is an infinite, omnipresent, three-dimensional being that is distinct from any particular body or the sum of all bodies.

Few thinkers were willing to take the divinization of space this far. Instead of saying God is space in the strong sense of identity, most preferred to say that space is an attribute of God or an effect of God.

This omnipresent void is dependent on God. If the void were destroyed when an object enters it, then the coincidence between God and space would eventually be broken by creation. However, Augustine bequeathed a conception of space as being capable of complete penetration by a material object. Under Augustine's accommodating conception, God and space perfectly coincide. Just as a statue coincides with the clay that composes it, God can coincide with space and space can coincide with objects that happen to occupy it.

Augustine's conception of space explains God's freedom. God could have created the world any*where* he wished. The whole universe could have been a cubit to the left of where it is currently situated. Not all location is relative to other bodies. Moreover, God could slide the whole universe to this location. So there is also *absolute* motion.

Instead of making our cosmos, God could have made another cosmos far away. Or God could have made both cosmoses or yet more cosmoses. Only by revelation do we know that God made a single cosmos.

Bradwardine's Five Corollaries

Bradwardine was a logician and mathematician. In *De causa Dei contra Pelagium*, Bradwardine rationally reconstructs Augustine's sponge as a system of five corollaries from the theorem that God is unchanging:

C1. God is everywhere in the world.
C2. And also beyond "in a place or imaginary infinite void."
C3. And can therefore be called immense and unlimited.
C4. Hence we can reply to the question: where was God before the Creation.
C5. Hence void can exist without body but not without God. (Grant 1981, 135–41)

Corollary C1 follows from God's immutability. If he were in one place and not another, then God would have to travel to exert his power. Travel implies change. Bradwardine presupposes there is no action at a distance. Not even God can act through the barrier of nothingness (the Great Insulator).

Some objected to locating God in vile places such as latrines. More abstractly, the sinners in hell (at the center of the earth) are supposed to be separated from God. By this geographical logic, God cannot be in hell. If God is anywhere, he must be in the best place, up in the heavens. He acts remotely by fiat rather than by direct contact.

The medievals inherit Aristotle's intuition that some directions are better than others. Up is better than down. Right is better than left.

Corollary C2 alludes to the space defined by our faculty of imagination. This space was acknowledged by the Aristotelians but characterized as fictive (fig. 16.4). The middle arch is inconsistently portrayed. (The miniature dates from before 1025. It is from a book of Henry II kept in Munich library.) Few viewers notice the inconsistency. The artist may have tolerated the optical impossibility as an insignificant technicality.

We are able to visually represent scenes that are spatially incoherent. There is no comprehensive consistency filter for perception and imagination. Since reality is consistent and imaginary space is not, the Aristotelians distinguished imaginary space from objective space.

Fictive space can be useful for certain purposes. People do navigate around their towns with their mental maps (which are rendered inconsistent by repeated applications of the heuristic that street corners

Figure 16.4 Impossible arch

form right angles). The Aristotelians did not condemn cartographers who made flat maps in which one travels off the right edge only to reappear on the extreme left side.

The ban of 1277 required Christians to regard some forms of imaginary space as at least logically possible. This commitment to the consistency of some forms of imaginary space can be vindicated by the mathematical utility of some forms of distorted space. Consider an angel on a cubical box who wishes to walk to the opposite side (fig. 16.5).

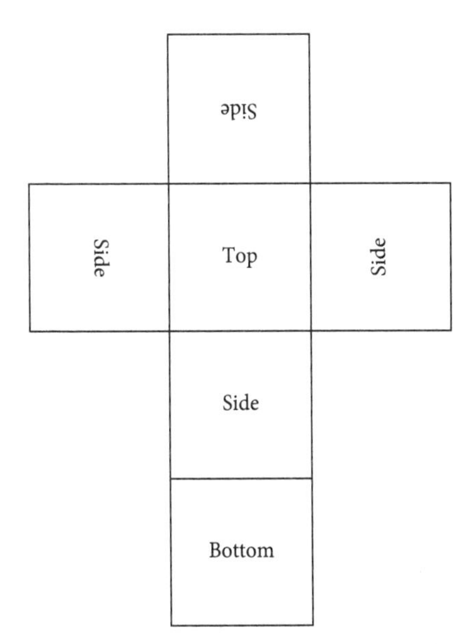

Figure 16.5 Unfolded cube

The angel's shortest path can be calculated by unfolding the cube into a two-dimensional plane containing the faces of the cubes as cells.

Once we learn the trick of compressing three-dimensional space into two-dimensional space, we are poised to expand from three-dimensional space to four-dimensional space. In the nineteenth century, Reverend Edwin Abbott (1838–1926) accomplished with his pop-out tale of Flatland. Just as two-dimensional creatures would find three-dimensional reality hard to imagine, we three-dimensional creatures find a four-dimensional reality difficult to grasp. Yet even if we cannot visualize it, we can understand with algebraic methods. A two-dimensional world is accessible with a coordinate system employing (x, y), a three-dimensional world with (x, y, z), and a four-dimensional system with . . . Heavens! I have reached the conceptual limit of my alphabet!

Some of Abbott's contemporaries believed that heaven was in the fourth dimension. The spiritualists among them claimed that residents of this fourth dimension signaled their presence by feats such as turning a right shoe into a left shoe by flipping it through the fourth dimension—just as we three-dimensional beings could amaze two-dimensional Flatlanders by flipping a *b* into a *d*. Saint Paul's Letter to the Ephesians contains the following four-dimensional passage: "You, being rooted and grounded in love, may be able to comprehend with all the saints what is the width and length and depth and height" (Ephesians 3:17–18).

All sides agreed that this imaginary space excludes a boundary. Corollary C2, that God also dwells in an imaginary infinite void, alludes to the space not allowing an end and therefore being infinite. Bradwardine is affirming the possibility of extracosmic void (contrary to Aristotle).

Jean Buridan had maintained that space depends on bodies—for it is nothing but a dimension of a body. If you were at the boundary of space and extended your arm, your arm would alter the boundary of space. Just as your hand makes space when extended between sofa cushions, your hand makes space when you extend it past the (former) boundary of space.

If God made a bean in otherwise empty space, then that bean would carry its own space. The distance between the isolated bean and everything else would be incommensurable.

On Buridan's theory of "internal space," there can be space between objects only if there are intermediate objects to host that space. Given that space is an aspect body, empty space is no more possible than a mountain without a valley. Bradwardine is opposing Buridan.

Corollary C3, that God is immense and unlimited, is inspired by a twelfth-century *Book of the XXIV Philosophers*: "God is an infinite sphere whose center is everywhere and circumference nowhere." Whereas the atomists infer that an infinite space has no center, the Christians inferred that the center is everywhere.

Thanks to the infinite cosmic void, the amount of space in any direction is the same. So space has a key feature of a sphere. This equality holds regardless of one's point of origin. Since the center can be anywhere, it is everywhere. Finally, there is no outer boundary of an infinite space, so the infinite sphere lacks a circumference.

Corollary 4 answers "Where was God before Creation?" with the precreation void. God is everywhere and every when. So he existed before Creation. Since the void depends only on God, the void is older than any body.

The imaginary void is not a *created* actual infinite. It is an aspect of the only being that is an actual infinite: God.

At first glance, Corollary 5 is a revival of atomism because the void no longer depends on bodies. Bodies can be subtracted but the void cannot. However, Bradwardine does make the void dependent on God.

Spinoza would later explain this dependence as space being the internal space of God, thereby allowing that God has a three-dimensional form. In contrast, Bradwardine's God has no dimensions. Consequently, the space in which God is situated is equally without a dimension.

God comes first. His freedom takes precedence over human freedom. His freedom enslaves space and time. The void emerges as a beneficiary. But the void must contort to fit its Master. The void must stretch beyond three dimensions to accommodate the possibility of turning left-oriented objects into right-oriented objects. It must shrink to accommodate the possibility of a Flatland, or a Lineland, and even become dimensionless to house God himself.

17
Newton
A Safe Space for Absence

Christians have a love-hate relation with atomism. They hate the atoms. They love the void.

Hate came first. As Epicurus emphasized, combinatorial explanations are reductive. Armies are *nothing but* soldiers. Soldiers are *nothing but* atoms. Atoms are *nothing but* indivisible bits of matter. Atoms leave nothing for the gods to do. Since we should postulate only what explains, atomists vacuum away the supernatural. Yet, as crystallized by the ban of 1277, Christians were obliged to overcome Aristotle's objections to the possibility of a vacuum. Awkwardly, only atomism affords empty space an explanatory role.

Space as Religious Object

Vast spaces are awesome. In Asia, these feelings of sublime spaciousness mushroomed into sky worship. The Mongols rhapsodized about how the sky is omnipresent and envelopes everything. Yet the sky is distinct from any of the things in the sky—all of which could be removed without removing the sky. The sky is indestructible. The sky cannot fall. The sky is eternal.

Arghun Khan sent letters to Philip the Fair in 1289. The documents contained ritualistic appeal to the divine blue sky. The Great Khan followed this with a policy letter in 1290 addressed to Pope Nicholas IV, conveying the religious pluralism and tolerance of the Mongols.

Although the Christian worships neither sky nor space, Thomas Bradwardine demonstrated how space bears an intimate connection with God. Only infinite space gives God a sufficiently capacious arena for action. A related motive for infinite space is God's omnipresence. If space were limited, God would inherit that limitation.

If God created everything, did he also create empty space? There are traditions that explicitly characterize nothingness as a divine artifact. According to the Jewish Kabbalah, God began with only himself. He emptied himself to make space for otherness. Contemporary Kabbalahists hail Big Bang cosmology as a decipherment of their mystical doctrine. God contracts himself to a point and, in a free act of self-sacrifice, explodes into space created by his evacuation.

Atomism evacuates occupants more methodically. The principle of parsimony instructs the atomist to substitute void for atoms wherever possible. New discoveries about how to swap the void for atoms yields a less populated universe.

If space could be shown to be suitably spiritual and dominant, along the lines paced out by Thomas Bradwardine, Christianity could be reconciled with atomism. Perhaps the atoms themselves could be rendered less mechanical by having them operate by forces. The atoms might even be spiritualized. Instead of viewing them as equal partners in a matter/void duality, one could reduce atoms to the void. Perhaps rocks are just regions of space endowed with properties such as impenetrability, color, and sound.

If these spiritualized atoms explained everything, then Christians could celebrate the comprehensiveness of space. If mysteries remained, these could be taken as evidence of divine intervention. Either way, Christianity wins.

Atomism was worth a second look. Inconveniently, Christian persecution of atomists had left Greek atomism in an unmarked grave—forgotten even by the gravediggers.

The Continental Resurrection of Atomism

In 1417 the bibliophile Poggio Bracciolini exhumed a copy of Lucretius's *On the Nature of Things* from a German monastery. Poggio was the former apostolic secretary to several popes. This literary corpse was slowly passed from the antiquarian bibliophiles to the new natural philosophers. The poem emerged, as a book, in 1474.

In France, Father Pierre Gassendi (1592–1655) led the theoretical resurrection of atomism. He claimed that God had created the atoms.

The post-1277 thought experiments that merely tried to show the possibility of a vacuum gave way to experiments employing pumps to evacuate air. *What if* became *what is*. The pious Blaise Pascal (1623–1662) designed the most theoretically well-grounded experiments on the vacuum. This led to Torricelli's barometer.

We live at the bottom of an ocean of air. A balloon is *pushed* up because its airbag displaces a large volume of air. The gas in the balloon has weight. This reduces lift. To increase lift, make a rigid balloon and remove the contents. In 1670, the Italian monk Francesco Lana de Terzi proposed that a gondola be attached to these maximally efficient balloons (fig. 17.1).

Sadly, seventeenth-century shipwrights lacked building material for the shell containing the vacuum. If the shell were strong enough to avoid imploding, its weight would swamp the extra lift.

As of 2021, science still lacks this strong material for the shell. Science fiction has possessed the substance since 1930: Harbenite. In *Tarzan at the Earth's Core* the king of the jungle travels to Pellucidar in a vacuum airship constructed of this non-break-through material.

Perhaps chemists should switch their focus from strength to homogeneity. If the shell were uniformly spherical, then even an arbitrarily thin balloon would work. There would be no more reason for the balloon to implode at one point rather than another. The absence of matter would be protected by the absence of reason!

Personalizing Nothingness

After a mystical experience in 1654, Blaise Pascal's interest in nothingness graduated from its significance to physics to its significance to people. Human beings have a unique perspective on their finitude. Pascal's *Pensées* is a rollercoaster ride surveying the human lot. Pascal elevates us to the level of angels by exalting our grasp of the infinite, and then submerges us below the beasts for wittingly choosing evil over goodness. From this canyon of depravity Pascal pulls us up again by marveling at how human beings tower over the microscopic kingdom,

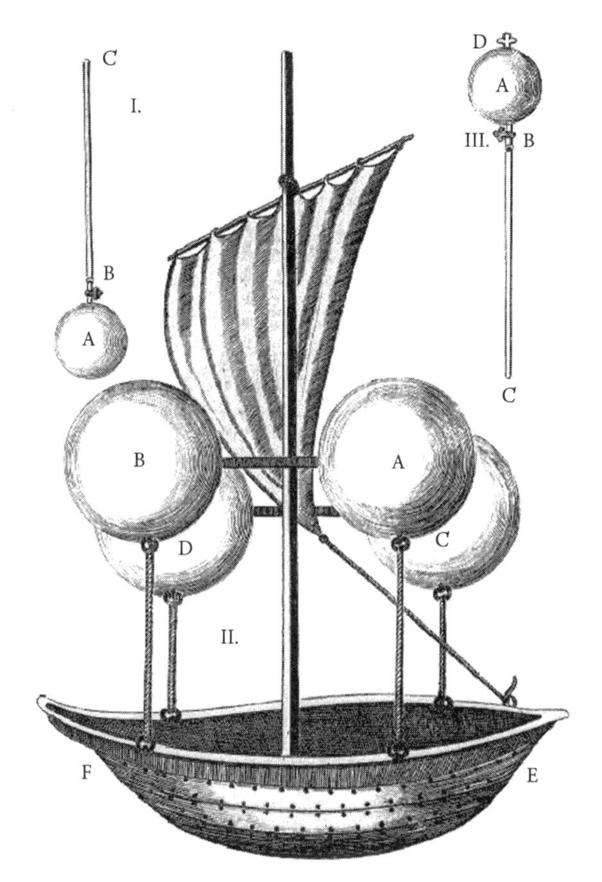

Figure 17.1 Vacuum gondola

only to plunge us down toward insignificance by having us dwell on the vastness of space, and the immensity of eternity.

> He who regards himself in this light will be afraid of himself, and observing himself sustained in the body given him by nature between those two abysses of the Infinite and Nothing, will tremble at the sight of these marvels; and I think that, as his curiosity changes into admiration, he will be more disposed to contemplate them in silence than to examine them with presumption.

> For in fact what is man in nature? A Nothing in comparison with the Infinite, an All in comparison with the Nothing, a mean between nothing and everything. Since he is infinitely removed from comprehending the extremes, the end of things and their beginning are hopelessly hidden from him in an impenetrable secret; he is equally incapable of seeing the Nothing from which he was made, and the Infinite in which he is swallowed up. ([1669] 1910, 72)

Pascal's association of nothingness with insignificance and meaninglessness broaches themes that would be elaborated by existentialists in the nineteenth and twentieth centuries.

The vacuum was brought down to earth by spectacular demonstrations. In 1654, Otto von Guericke (1601–1686) attached two copper hemispheres with a mating rim, pumped the air out, and had a team of 30 horses try, in vain, to pull apart the hemispheres. When Otto opened a valve to let the air in, the spheres separated without any pulling. (The pump and spheres are on exhibit at the Deutsches Museum, Munich, Germany. Otto von Guericke went on to demonstrate the elasticity of the air, air pressure, and undermined the principle that nature abhors a vacuum.)

Fridugisus had celebrated nothingness as the ultimate building material. Otto von Guericke comes close to identifying nothing with the Builder:

> Everything is in Nothing and if God should reduce the fabric of the world, which he created, into Nothing, nothing would remain of its place other than Nothing (just as it was before the creation of the world), that is, the Uncreated. For the uncreated is that whose beginning does not preexist. Nothing contains all things. It is more precious than gold, without beginning and end, more joyous than the perception of bountiful light, more noble than the blood of kings, comparable to the heavens, higher than the stars, more powerful than a stroke of lightning, perfect and blessed in every way. Nothing always inspires. Where nothing is, there ceases the jurisdiction of all kings. Nothing is without any mischief. According to Job the Earth is suspended over Nothing. Nothing is outside the world. Nothing is

> everywhere. They say the vacuum is Nothing; and they say that imaginary space—and space itself—is Nothing. (Guericke 1672, 63, as translated by Grant 1982, 216)

The British Resurrection of Atomism

In England, the revival of atomism had a false start. In the late sixteenth century, the Ninth Earl of Northumberland, Henry Percy (1564–1632), supported a group of natural philosophers who were all Epicurean atomists with the single exception of Nathaniel Toporley. In a letter describing research on optics, Thomas Hariot invited Johannes Kepler to become an atomist:

> I have now led you to the doors of nature's house, wherein lies its mysteries. If you cannot enter because the doors are too narrow, then abstract and contract yourself into an atom, and you will enter easily. And when you later come out again, tell me what wonders you saw. (Hariot to Kepler, in Kepler, *Werke*, 15:368)

The group was secretive because of the hostility to atheists. But the conspirators were not secretive enough. As reputed atheists, they became magnets for imaginary immoralities, most notably, a plot to assassinate the king. The "Wizard Earl" Percy was fined. And imprisoned. His group—disbanded.

This shattering setback suggested a need to dissociate atomism from Epicurus. These efforts to Christianize atomism were complicated by impolitic atomists such as Thomas Hobbes (prior to becoming a Plenist) with his materialist theology. The "mad Duchess" Lady Margaret Cavendish also alarmed this second generation of atomic resurrectionists.

Atomists turned to Christianizing propaganda. The professor of Hebrew Ralph Cudworth (1617–1688) created a biblical lineage for atomism. He traced atoms back to a Phoenician—who turned out to be Moses! As for Epicurus, he was a plagiarist. He perverted Old Testament metaphysics into a premise for atheism. Christians should reclaim atomism as their own!

The British natural philosophers were led by the theoretician Walter Charleton (1619–1707). Whereas Epicurus conceived of the atoms as a purposeless swirl of atoms that fortuitously congregates into birds and bees, Charleton saw the complexity of the objects as further evidence for the argument from design. Who but God could direct the many atoms into exquisitely well-functioning organisms?

On the experimental side, Robert Boyle (1627–1691) was the leader. In addition to performing his own experiments, the wealthy Boyle subsidized Robert Hooke's technical advances with the vacuum pump.

Absent-Minded

There are many anecdotes about Isaac Newton's absorption in his work. His maid spotted him in his kitchen, holding an egg and boiling his watch. His nephew reports, "At some seldom times when he designed to dine in the hall, would turn to the left hand [rather than going straight], and go out into the street, where making a stop, when he found his mistake, he would hastily turn back & and then sometimes instead of going into hall, return to his chamber again" (Brewster 1855, 96).

Thomas Moore's diary relates: "Anecdote of Newton, showing his extreme absence—inviting a friend to dinner, & forgetting it—the friend arriving, & finding the philosopher in a fit of abstraction—Dinner brought up for one—the friend (without disturbing Newton) sitting down & dispatching it, and Newton, after recovering from his reverie, looking at the empty dishes & saying, 'Well really, if it wasn't for the proof before my eyes, I could have sworn that I had not yet dined'" (Moore et. al. 1983, vol. 3, 1104).

As a fan of absences, my favorite tale is of Newton's kitten hole. His cat already had a hole in the door through which it could pass. After the cat gave birth, Newton made a smaller hole for the kittens.

Newton's absent-mindedness was a side-effect of his single-mindedness. When his attention turned to making money, as Master of the Royal Mint, he literally made better money. Newton's successful re-coinage of 1690 was coupled with zealous prosecution of counterfeiters. To increase his quantity of this better quality money, the Master of the Mint became a speculative investor. What mattered was the price subsequent investors would be willing to pay for the asset, not the objective value. As long as there was some greater fool willing to pay an even higher price, the objectively over-priced asset was a lucrative investment. Just do not be the last fool! This marketplace of iterated expectations made Newton very rich. He remained rich even after he lost a fortune in the South Sea Bubble of 1720. Newton is widely quoted as sadly concluding that "he could calculate the motions of heavenly bodies, but not the madness of people" (Odlyzko 2018, 29).

Increasing the Share of Empty Space

While a student at Cambridge University, Newton (1642–1727) kept a "philosophical" notebook. It shows that he was first exposed to Epicurean atomism around 1664 through Walter Charleton's 1654 *Physiologia Epicuro-Gassendo-Charltoniana: A Fabrick of Science Natural upon the Hypothesis of Atoms Founded by Epicurus, Repaired by Petrus Gassendus, and Augmented by Walter Charleton*. Following the first-generation resurrectionist Pierre Gassendi, Newton gives God primacy over the atoms. At first there is an infinite empty space that has an infinite past. Then about six thousand years ago, God created the atoms and directed their movements for his ends. Instead of interacting in a purely mechanical manner, by virtue of the shapes and sizes, Newton introduces forces. These residues of alchemy operate between atoms in a symmetric fashion. Given this symmetry and the absence of mechanical interactions, Newton tends to assume atoms are uniform spheres.

In addition to local forces, Newton sheepishly adds the alchemical universal force of gravity. Later theorists would dematerialize the atom by having forces do more and more of the work formerly assigned to mechanical features of the atoms.

Atomists have always welcomed arguments that increase the ratio of empty space to matter. In the *Optics*, Newton contributes a hierarchical argument. Historically, units of matter have been revealed in stages as our ability to divide has improved. At each of these stages, what appeared to be solid was revealed to be porous. Eventually, there must be a final stage at which all the particles are genuinely without pores. But if there were n intermediate stages, the amount of empty space has increased n-fold—an exponential rate of growth. Familiar objects around us, therefore, are mostly empty space.

Newton's reasoning has been vindicated. If all of the matter that comprises your body were swept together, you would be a speck. Others may marvel that you are mostly water. Newton marvels that water is mostly empty space.

The Priority of Space

René Descartes takes extension to be the essence of bodies and so precludes vacuums by definition. In *De Gravitatione* Newton counters that impenetrability is crucial to the concept of a body.

Newton underscores this with a divine recipe. Given any body, God could create a duplicate comprised of regions endowed with special properties. To make a "door," God specifies some door-sized volume. He then endows this region with the property of impenetrability (a kind of repulsive force). You cannot walk through the "door." The "door" is impenetrable to light. Thus you can see it by virtue of the light it reflects. The "door" is also impenetrable to sound. So you can hear it by virtue of the sound it reflects. Lastly, the "door" is mobile; it can travel through space like wave. Adjoining regions inherit the impenetrability of the door in a lawful way (specifically Newton's laws of motion). The "door" is a solidified shadow. Swing the "door" open. Tell me what wonders you see.

One may take this "door" as a phantom—the doorway is open despite the fact that is something that looks like a door and sounds like a door and blocks you like a door. Or one might remove the scare quotes and conclude that a door is nothing but regions of space peppered with properties.

Divinity of Space

For Newton, space is a touchstone of reality. Anything real must either be space or be related to space. In his unpublished, pre-*Principia* treatise *De Gravitatione*, Newton writes:

> Space is an affection of a being just as a being. No being exists or can exist which is not related to space in some way. God is everywhere, created minds are somewhere, and body is in the space that it occupies; and whatever is neither everywhere nor anywhere does not exist. And hence it follows that space is an emanative effect of the first existing being, for if any being whatsoever is posited, space is posited. . . . If ever space had not existed, God at that time would have been nowhere. (2004, 25)

Since God is real, he must relate to space. God cannot be outside of space because he is omnipresent. He cannot be in space because then space would limit him. So space must be an emanative effect of God.

More specifically, space is God's immensity just as time is God's eternality. Does this mean that God has parts? This would threaten God's unity and simplicity.

Objects with parts are divisible. But the parts of space are unified in that they cannot be pulled apart from each other. Each part of space relates to the remaining places as numbers relate to other numbers on the number line.

Space is also God's sensorium—a sense organ by which God knows everything. Since each thing is in space, God immediately perceives it. God "in whom we live and move and have our being" (Acts 17:28) knows everyone and everything in the sense of having direct acquaintance with it. If your wife has an identical twin whom you have never met, then you know your wife's body in the direct "biblical sense" while only knowing her twin's body indirectly. God knows us all directly in the biblical sense.

Isaac Newton may have been influenced by his friend's theory of property. According to John Locke, you come to own objects by mixing your labor with them. When you whittle a stick into a cane, the cane becomes yours. When you catch and tame a wild ass, the capture and taming transforms the creature into your ass. Analogously, your right leg becomes your leg by mixing sensations with it. Projecting a phantom leg into a prosthesis improves your ability to perambulate with it. The "leg" becomes part of your body. If you could control the arm attached to my shoulder in the same way, that arm would be part of *your* body, not mine. If those fingers fondle my ears, I shall justly object to your inappropriate touching.

Newton knew that John Locke analyzed personal identity as continuity of consciousness. If a prince's consciousness takes control of a cobbler's body, then that body becomes a full body prosthesis for the prince. The same holds for the cobbler. The pair could switch bodies in the same way two farmers could switch pastures.

Under this sensational account of body ownership, God's omnipresent sensitivity makes all of space his body. This tactile sensitivity makes God's body differ from Plato's perfect animal. In the *Timaeus* Plato describes the world as an organism (round like a pufferfish but with a perfectly smooth surface). Creatures within this perfect animal have sense organs because they need to learn about their environment. But the perfect animal is all-encompassing. Since there is no environment, there is no need for organs to sense (or mate or excrete). Superfluous organs would be a disorder, and the perfect animal is free of diseases. The perfect animal thinks but does not perceive.

Newton would object that the senses are not restricted to sensing what is outside. When you rotate, you feel centripetal force. Since Plato's perfect animal is a rotating ball, it could have a sense of balance that would detect its rate of spin. When your muscles contract, you sense-perceive this internal change. Plato's animal has many internal parts that could be monitored kinesthetically.

Since God is a necessary being, space must also be a necessary being. Yet space still depends on God. Consider God's relationship to his singleton set: {God}. Although that set necessarily exists, God is the more fundamental entity, just as {Sun} depends on the Sun and not the other way around.

Newton's Ambivalence

Is space a substance or a property? Newton thinks neither answer quite fits—though space comes closer to qualifying as a substance than as a property.

A substance must be capable of acting on any substance. Space does not cause anything. Mere difference in spatial location cannot account for any difference between two objects. We do speak of mountains wearing down with time. But when pressed, we transfer responsibility to processes operating within time. Time itself cannot wear down a mountain—or do anything.

Space cannot be a property because properties depend on substances. But empty space is possible.

Yet Newton carefully adds: "Much less may [space] be said to be nothing, since it is something more than an accident, and approaches more nearly to the nature (naturam) of substance" (2004, 22).

Initially, Isaac Newton's space was majestically empty. This gratified theologians. They recalled Augustine's metaphor of the material world being a sponge in the sea. Newton agreed with the theologians that space does not meet the Eleatic criterion of reality: causal power. So space is not a clear case of a substance. And space is definitely not a property; all properties depend on substances, and space existed prior to there being any objects. But Newton refuses to conclude that space is nothing. Ultimately, Newton disappointed the theologians by filling space with ether (to compensate for the causal inertness of space). Newton was ambivalent about the addition; he feared friction with the ether would slow the orbit of the planets, ultimately causing everything to STOP.

18

Leibniz

Absence of Contradiction

Gottfried Leibniz (1646–1716) rejects the Christian resurrection of the void. God is the opposite of nothingness. Being flows from God as water from a fountain. The water overfills the basin, filling the universe.

Possible ∴ Actual

God draws himself into existence by his own nature (fig. 18.1).

To see how, consider Seneca's definition of 'God': that which nothing greater can be thought. Seneca's unit of measurement was physical magnitude. Saint Anselm notes that the *fundamental* measure of greatness is goodness. Accordingly, God is the best conceivable being. This definition elegantly answers a catechism about God's nature: What does the best conceivable being know? *Everything.* How powerful is the perfect being? *All-powerful.* How good? *Perfect.* Having concisely established the nature of God, Anselm applies Seneca's definition to the question of God's existence. Which would be the greater being, one that exists or one that does not? Being beats nonbeing! Anselm concludes that we must think of the greatest conceivable being as existing.

Gottfried Leibniz is an optimist about proving God's existence: "Almost all the methods which have been used to prove the existence of God are sound, and could serve the purpose if they were rendered complete" (1996, 9). Leibniz proceeds to fill the gaps in the proofs of the Divine Gap-Filler.

The key incompleteness of the ontological argument is the absence of a consistency proof for the definition. The need for such a proof can be illustrated with Oskar Perron's "proof" that 1 is the largest

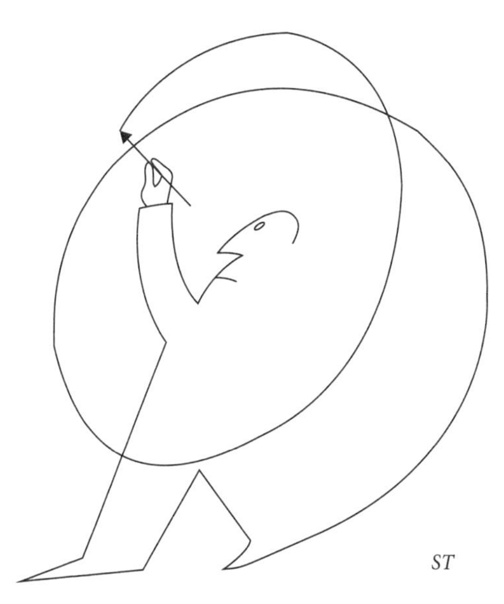

Figure 18.1 Saul Steinberg self-creation

natural number. First, define N as the largest natural number. N^2 cannot be greater than N because N is defined as the largest number. So, $N(N - 1) = N^2 - N$ is not positive. Therefore, $N - 1$ is not positive. Accordingly, N cannot exceed 1. Yet N is at least 1. It follows that $N = 1$. Since N was defined as the largest number, 1 is the largest number. Perron's proof is valid; the premises do entail the conclusion. But the proof has a false premise because 'the largest natural number' is an inconsistent description. The concept of a natural number requires that each number have a successor. Each step of Perron's deduction is valid. However, the whole proof is trivial. We are merely drawing out consequences from a contradiction. Anything follows from a contradiction. So the worry is that the ontological argument shares the trivial validity of: *God is not God; therefore, God exists.*

A proof of possibility in mathematics is a far more powerful result than a proof of possibility in physics. For in mathematics, there is no difference between proving that something is possible and proving that it is actual. If it is possible that there is an odd perfect number, then there is such a number. And if that number actually exists, it necessarily exists. (In mathematics, possibility, actuality, and necessity converge: anything possible is actual, anything actual is necessary, so anything possible is necessary.)

A perfect number is any number that equals the sum of its proper divisors: 6 is the first perfect number because $1 + 2 + 3 = 6$. The second perfect number is $28 = 1 + 2 + 4 + 7 + 14$. In a 1638 letter from René Descartes to Marin Mersenne, Descartes conjectured that all even perfect numbers can be generated from a formula given by Euclid. Descartes then added that he did not know of anything that precludes there being an odd perfect number. If there really isn't anything stopping it, then should we expect an odd perfect number somewhere along the infinite sequence of numbers?

Mathematicians still do not know whether there are any odd perfect numbers. They know that any such number must be larger than 10^{1500} . Its largest prime factor must exceed 10^8. They know that any odd perfect number must be a number divisible by 15. Centuries of profiling have given mathematicians much *conditional* knowledge of odd perfect numbers.

Centuries of theology have done the same for God. This explains why Leibniz labored to prove that the definition of 'God' harbors no hidden contradiction. Since Leibniz thought his consistency proof succeeded, he proceeded to the next stage of his metaphysics.

Why Is There Something Rather Than Nothing?

The number 1 cannot exist in isolation. To be a natural number, 1 must have a successor. Since any successor of a number is itself a number, there must be infinitely many numbers.

A solitary god would not be as great as a god who brought other things into existence. And it is best to bring about the best—presumably another god.

However, this creative path from monotheism to polytheism is blocked by a pair of logical principles. First, the law of contradiction says that nothing entailing a contradiction can exist. If there were two all-powerful beings, one could be immovable while the second was irresistible. But that would lead to a contradiction of the immovable both moving and not moving. An immoveable object is possible and an irresistible force is possible. But they are not co-possible. Any creation must be congruent with the rest of what exists.

Second, the principle of sufficient reason forbids duplication of gods (or anything else). If there were two perfect beings, there would be no reason why one god is distinct from the other god. There must be a *sufficient* reason for each truth.

I might try to explain the contents of a geometry book by the fact that it is a copy of an earlier geometry book. I explain the earlier book by it being a copy of a yet earlier book. Insufficient reasons are not enough, even when there are infinitely many of them.

The principle of sufficient reason extends to truths about what does not exist. If there are no unicorns, then there must be a reason for their absence. Since there was no such reason, Leibniz expected unicorns (and monopoles, and microscopic organisms). Of course, positive evidence is stronger than a mere presumption of existence and Leibniz had some: Mayor Otto von Guericke's reported that unicorn bones were being bought near his city of Magdeburg. In 1686, Leibniz traced the unicornucopia to a cave in the Harz Mountains (now dubbed "Unicorn Cave" by tour operators). Leibniz did a paleontological reconstruction based on skeletal remains (fig. 18.2).

Perhaps Leibniz is also reconstructing an anecdote. According to Diogenes Laertius, Plato defined man as a featherless biped. Diogenes the Cynic plucked a chicken and brought it to Plato's Academy: "Behold! I have brought you a man!" Leibniz's sense of humor drew a compliment from the Duchess of Orleans: "It's so rare for intellectuals to be smartly dressed, and not to smell, and to understand jokes." Leibniz would have been amused to learn that the original featherless bipeds in Unicorn Cave were Neanderthals who arrived before *homo sapiens*: "A 51,000-year-old engraved bone reveals Neanderthal's capacity for symbolic behavior" (Leder et. al. 2021).

Although previous rationalists have slogans that sound like the principle of sufficient reason, they normally limit the scope of explanation. In the *Timaeus* Plato writes, "Everything that comes to be must of necessity come to be by the agency of some cause, for it is impossible for anything to come to be without a cause" (28a4–5). But in Plato's creation story we discover that there are things that are not among

Figure 18.2 Leibniz's paleontological reconstruction of a unicorn in his *Protagaea*

the things that "come to be." The Demiurge imposes order on *preexisting* chaos. This disorder is uncaused. This sort of brute beginning is presupposed by almost all creation stories.

Given that each thing has a reason, there must be a reason for everything. For the conjunction of all the truths is itself a truth. Consequently, there must be a reason why the actual world exists rather than some other possible world. The only candidate for such a complete explainer would be God. So, the principle of sufficient reason provides another logical proof of God's existence.

To learn the nature of the actual world we need to understand God's basis for making decisions. Since God always acts in a principled fashion, we have a powerful clue about what God can construct. Just as the law of contradiction precludes God from making a cubical sphere, the principle of sufficient reason precludes God from making a solitary sphere. There would be no more reason to make that sphere one size rather than another. (In contrast, God can connect point A to point B with a line; there is a supremely efficient solution: draw a straight line between A and B.)

We constantly use the principle of sufficient reason in our inferences. Consider Archimedes's reasoning: a scale with equal weights must be balanced because there is no more reason for one side to ascend rather than the other.

If there were exact twins, then God would have no more reason to place one twin to the left rather than right of the other twin. This arbitrariness would be incompatible with God's divinity. Consequently, distinct things must differ in their intrinsic properties. More positively, if x and y have the same properties, then x = y. This is Leibniz's principle of the identity of indiscernibles.

Leibniz complains that empiricism is erected on a violation of the principle. According to the founder of empiricism, John Locke, we begin as blank minds. In this sense, all men are created equal: 0 = 0 = 0. . . . But the degenerate nature of this equality is revealed by the principle of indiscernibles. If there is no difference in minds, there is only a single mind.

Each person must begin with ideas. Furthermore, the set of innate ideas must be unique to that individual. To be is to think. And to think is to think *differently*.

Just as it is good for the creator to maximize the number of substances, it is also good to maximize the number of instantiated properties. Since properties can only be instantiated by substances, the maximization of substances promotes the maximization of properties.

Perfection is harmony. Harmony is unity within variety. Therefore, the actual world contains the richest phenomena governed by the simplest of laws. The creation of each being must be coordinated with the creation of every other being. If there were a tie for the maximal combination of simplicity and variety, then God would not have created any world. Since there is a world, there is a collection of laws that most efficiently yields the richest phenomena.

From God's perspective, the world has the completeness of the number line. Since God cannot make duplicates, he will make

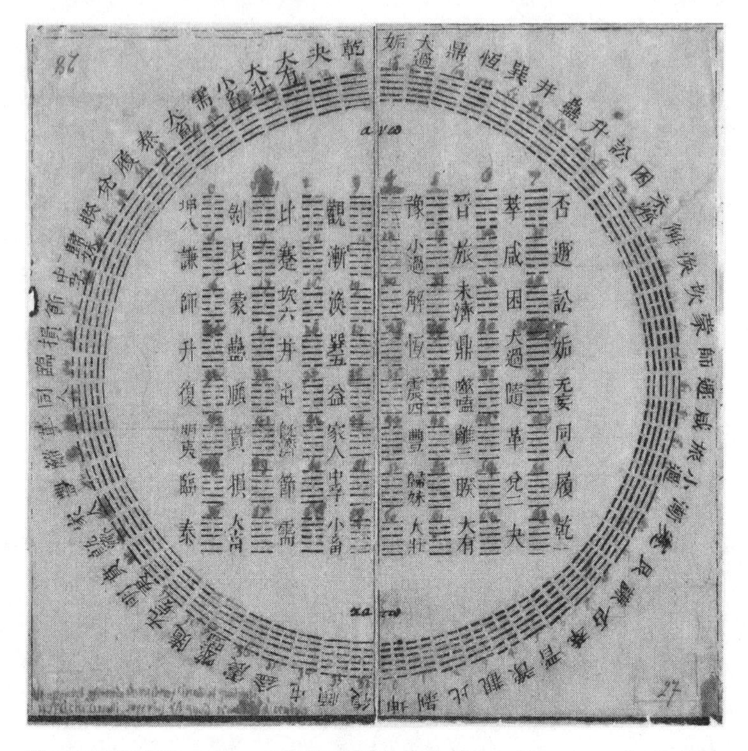

Figure 18.3 Diagram sent by Bouvet to Leibniz

one second best being, *one* third best . . . There can be no gaps in the chain of being. Each link is essentially the successor of its predecessor.

Confirmation of Christianity from China

Leibniz was among the first to develop binary notation. This is the simplest system; it only requires the duality of two symbols. In 1701 the Jesuit missionary Joachim Bouvet sent a diagram suggesting that the ancient Chinese used binary notation (fig. 18.3).

The hexagrams from the *Book of Changes* correspond to binary notation (fig. 18.4).

<table>
<tr><td>¦¦¦</td><td>000</td><td>0</td><td>0</td></tr>
<tr><td>¦¦|</td><td>001</td><td>1</td><td>1</td></tr>
<tr><td>¦|¦</td><td>010</td><td>10</td><td>2</td></tr>
<tr><td>¦||</td><td>011</td><td>11</td><td>3</td></tr>
<tr><td>|¦¦</td><td>100</td><td>100</td><td>4</td></tr>
<tr><td>|¦|</td><td>101</td><td>101</td><td>5</td></tr>
<tr><td>||¦</td><td>110</td><td>110</td><td>6</td></tr>
<tr><td>|||</td><td>111</td><td>111</td><td>7</td></tr>
</table>

Figure 18.4 Figure of the Eight Cova

Since no records could be found of binary arithmetic, Leibniz concluded that "the Chinese lost the meaning of the Cova or Lineations of Fuxi, perhaps more than a thousand years ago, and they have written commentaries on the subject in which they have sought I know not what far out meanings, so that their true explanation now has to come from Europeans" ([1703] 1859, 223). Leibniz took his discovery to illustrate the complementarity of Europe and China. The Chinese were wiser on practical matters. Almost all inventions originated in China. Thanks to its social engineering, China was more peaceful and more harmonious than Europe (which had just endured the Thirty Years War). The Chinese were also preeminent in applied fields such as geography and observational astronomy. Despite this superiority in practical reasoning, the Chinese were inferior in speculative arts and theoretical reasoning (the abstract geometry of Euclid, physics, and metaphysics). Chinese civilization had declined from a golden age in which they were better informed by the biblical prophets, such as the "ancient King and philosopher named Fuxi, who is believed to have lived more than 4000 years ago, and whom the Chinese regard as the founder of their empire and their sciences." (Fuxi is actually mythological figure, said to have lived in the third millennium BC.)

Christian missionaries could help unscramble Confucius's original truths. The Jesuit missionaries were not trying to *convert* the Chinese; the Jesuits were helping them recover their true beliefs.

One motive for this restorative theme is Leibniz's fear that the emperor of China would expel the Jesuits, just as the emperor of Japan excluded the Jesuits after 1596. The Japanese had initially welcomed the Spanish Jesuits to exchange ideas. The Spaniards were permitted to proselytize. But then a Spanish sea captain boasted that his countrymen conquered foreigners by first converting them. The Japanese Exclusion Edict of 1639 addressed this threat of a weaponized religion. Christianity was banned. Interaction with foreigners was severely restricted. Mindful of the Japanese reversal, the Jesuits in China did not wish to be perceived as the vanguard of a European invasion. They contended that Confucianism is compatible with Christianity. For instance, what appears to be ancestor worship is just a social rite with no more religious content beyond what is present in the Old Testament commandment "Honor thy father and thy mother."

This inaugurated a tradition of East-West parallels. In *The Tao of Physics*, Fritjof Capra treats parallels between Eastern religion and Western quantum mechanics as mutually corroborating. The ancient East and the modern West are independent witnesses whose consilience is best explained by them perceiving a common reality (in the way that the 1572 Milky Way supernova provides the best explanation of simultaneous reports of a "new star" by Tycho Brahe and Ming Dynasty astronomers).

In contrast to Capra, Leibniz posits a common *origin* of witnesses. Confucianism resembles Christianity because their chain of testimonies trace to the same source. All human beings are descended from Adam. Therefore, they all have the same capacity to discover religious truth through reasoning—natural theology. There may have also been, long ago, some *revealed* theology from followers of Moses. Admittedly, any testimony from the prophets were poorly preserved by the Chinese. Lamentable gaps in their annals are discernible from their failure to record the Deluge or the sun standing still for a whole day (Joshua 10:1–14). Chinese astronomers were also guilty of errors of commission, They report a "guest star" that would have been visible to the Europeans from July 4, 1054 to April 6, 1056. But there were no European records of this counterexample to the permanence of stars.

In the case of the hexagrams, Leibniz believed there was an opportunity to foster Chinese recollections of the creation story. He sympathized with Chinese resistance to interpreting Genesis as literal creation from nothing. But Leibniz thought that one aspect of the doctrine had value; the universe is perfectly designed. This emphasis on purposeful construction counters the Manichaean vision of the universe as rubble from the war between good and evil. Divine design is also an improvement over the Taoist idea of a universe arising from a mindless tension between yin and yang ☯.

Happily, Taoist verse gives a running start to the notion of creative progression: "Tao begets One, One begets Two, Two begets Three and Three begets all things." Leibniz believes that binary notation endows this progression with mathematical gravitas: "Now one can say that nothing in the world can better present and demonstrate this power than the origin of numbers, as it is presented here through the simple and unadorned presentation of One and Zero or Nothing." This is from Leibniz's letter to the Duke of Brunswick proposing that a coin be struck with this image.

In a response to Bouvet, Leibniz articulates the analogy arithmetically. The void corresponds to 0 and God to 1. The beginning is 0, then comes 1, God, who makes heaven and earth, 10. By the seventh day, 111, the world lacks 0, the void. Since everything has been made full, God rests. That is why the seventh day is declared the Sabbath. This supplements Saint Augustine's explanation of why creation took six days; six is the first perfect number (*City of God*, Part XI, Chapter 30).

Metaphysics as Reverse Engineering

When the ancient Egyptians disassembled a captured Assyrian chariot, they learned its structure, the function of its parts, its modes of operation. The Egyptians had to bear in mind that the chariot might be damaged. They had to bear in mind that there might be design flaws. No such caveats haunt the metaphysician when he reverse-engineers God's creation. For the world as a whole, the ideal and real exactly coincide.

Normally, the reverse engineer has the same mental powers as the original engineer. When a man designs a chariot, he minimizes the number of parts, the kinds of parts, and the amount of labor needed for assembly.

The survival of ancient Roman buildings is a testament to Roman ignorance. Their engineers erred on the side of safety. Contemporary buildings are lighter and less durable because engineers have a more exact understanding of how much is needed for a building that will endure a given number of years. Overbuilding is wasteful. In 1945 a B-25 Mitchell bomber crashed into the 79th floor of the Empire State Building—doing little damage. Embarrassed engineers frugally responded by building lighter, flimsier skyscrapers.

God is under no constraint concerning resources and labor. God can ascertain everything about a thing just by analyzing its nature. God's designs are apt to look risky and complicated. For example, a simple way *for a man* to tile a floor is to use uniformly colored squares. This is the most forgiving approach. The symmetry of each tile makes them interchangeable. God has no need to simplify. He has unlimited memory, attention, and foresight. God's preference for variety would lead him to employ tiles that are each unique in both shape and color pattern. Once assembled, the tiles would reveal novel patterns at larger scales. Thus, Roger Penrose's aperiodic tiling is closer to the divine than square tiling (fig. 18.5). Square tiles

Figure 18.5 Aperiodic tiles at the Mitchell Institute for Fundamental Physics and Astronomy, Texas A&M University

have translational symmetry; a copy of each region can be found by moving a fixed distance. Placed on a field of homogenous squares that extend to the horizon, you cannot tell where you are. Nothing changes as you walk. In aperiodic tiling, new patterns are found as you travel. Yet aperiodic tiling also has a pleasing self-similarity; patterns repeat at larger and larger scales. This conveys the idea of each region being a microcosm.

No Spatial Gaps

If God were limited to a finite number of tiles, he could not fill the space with round tiles. But if God has an infinite stock of round tiles, he could fill the interstices with smaller circles, and their smaller interstices with yet smaller circles. This scheme would satisfy God's preference for having infinitely many things.

However, a more efficient strategy would be to pack infinitely many sub-tiles into each tile. What might appear to us to be a large tile would actually be a composite of smaller tiles. Microscopic examination of the tiles would reveal that this principle of composition always holds: every tile is composed of a sub-tile.

Of course, there is a human limit to magnification. Finite experience cannot reveal an infinite nesting of tiles. But metaphysics can supply the missing information. The universe must be composed of gunk; entities whose parts always have proper parts.

Does the divine drive for variety mean that there are also some atoms mixed into the gunk for variety? No, atoms imply a contradiction. Each atom has a definite size and shape. Since atoms are finite in area, they must have a precise perimeter. Yet the perimeter of an object increases with the degree of resolution. If two countries measure their common border with different degrees of resolution, then they will assign different lengths to the shared border. Since there is no unique standard, atoms cannot have definite perimeters—a feature essential to their existence.

In addition to believing in gunk, Leibniz believes in knug—the well-named reverse of gunk. With knug, each thing is a proper part of something else. There is no largest object. So those who define 'world' as the largest object complain that the knug theory implies there is no actual world. They conclude 'junk' is a better name than 'knug'!

To maximize the number and variety of objects, the universe must be more than a two-dimensional plane. There must be as many dimensions as possible. Although the universe appears three-dimensional to us, there are other dimensions invisible to us.

Newton's allies addressed worries about wasted space with physics rather than metaphysics. Edmond Halley calculated that variations in earth's magnetic field are effects of a hollow earth. Below spin three concentric shells reminiscent of Saturn's rings. Their magnitudes are proportional to Venus, Mars, and Mercury. Explorations to the poles were partly motivated as searches for intraterrestrial intelligence. Success would have greeted as a return to the womb by Creek Indians. Their ancestors emerged from caverns near the Red River. A similar birth is claimed by tribes around the world.

No Temporal Gaps

Isaac Newton pictures God as waiting an infinite period before creating the world. Leibniz rejects this as a waste of time. Leibniz also objects to a future oblivion in which everything goes out of existence. In contrast, Newton believed in the doomsday prophesized in the Revelation.

These teleological bans on empty time and space are reinforced metaphysically by Leibniz's relational theory of space and time. Whereas Newton believes that space and time exist independently of objects, Leibniz defines space and time as relations between objects. He pictures space as an abstraction from relations between objects. Consequently, space can be described with the same metaphors we use for family trees. Space can grow. Space can curve, warp, or have holes. There is much room to wonder why space has properties that it has. Since space is an abstraction from objects, answers to any riddles about space reduce to facts about objects. One can wonder why there is space. But this is only to wonder why there are objects.

Leibniz welcomes change as an opportunity for more variety. But variety also requires that change be continuous. This provides another impossibility proof against atoms. The simplicity of atoms requires them to behave discretely during collisions. Any atomic change in direction must be instantaneous. Thus, an atom heading south before the collision must, at the moment of the collision, also head north. But nothing can head in opposite directions simultaneously.

And why think that atoms collide rather than merely pass through each other? Consider the negative interference of waves. When a northbound crest encounters a southbound trough, the water becomes level. Why not say the same of the atoms? Since there is no more reason to describe the situation as collision rather than interpenetration, neither description can be correct.

Democritus does not exclude atoms from being as big as an ox. But he talks as if they are too small to see. This obscures their peculiarities.

According to Leibniz, atoms are never small enough. However small they are, they could have been smaller and so have done the same work taking up less space. To avoid waste, the hierarchy of parthood must be bottomless.

Leibniz's gunk is more organic, and so more Chinese than Nagarjuna's gunk or Anaxagoras's gunk. Indeed, *all* Leibnizian gunk is living gunk. We are composed of organisms that are themselves composed of organisms. These suborganisms are subservient elements akin to organs. The superorganism organizes the hierarchy of suborganisms—and is itself organized by a yet higher-level organism.

Biologists presently picture the human body as an ecosystem in which alien bacterial cells (about a hundred trillion) live in a mostly cooperative fashion with your indigenous cells (about ten trillion). The bacteria are small but there is a lower limit. For Leibniz, there is no lower limit.

Nor are there limits on temperature or chemistry. The Leibnizian biologist expects extremophilic organisms. He searches for life everywhere and everywhen. Nothing is sterile.

In addition to believing in extraterrestrial life, Leibniz believed in extraterrestrial *intelligent* life. All organisms have a degree of mentality—and every degree of mentality has some organism.

If an omnipotent carpenter simply wished to fill space, he could pack the world with Albert of Saxony's cubes. To maximize variety, however, God must refrain from duplicates.

Any vacuum would contain smaller vacuums that would not differ in their intrinsic parts. So there can be a vacuum only if there are duplicate vacuums. Since the principle of sufficient reason precludes duplicates, the void is as impossible as the atom.

The criteria for God's choice of a world preclude any arbitrary element within the world. God chose a world that was simplest in laws and richest in phenomena. He was chiefly concerned with the happiness of minds—minds that deliver the maximum perspectives on the universe. So the world is a plenum, filled with minds each of which expresses the world from its own point of view. Any gaps are as illusory as the "gap" between 1 and 0.999 . . . (which are just distinct numerals for the same number).

Leibniz pictured possible things as competing to become actual. The more a thing competes with other things, the more likely that there will be something that stops it from becoming actual. Competitors lose in this snuggle for existence. The winners precisely fit the opening formed by other things. This niche is a keyhole into existence of all other things. The little bit that is not tells us about all that there is.

The actual world is the best *as a world*. But is it best for us? Perhaps the world best *for us* is an inferior world, say one with less variety (no mosquitoes). An evil that is necessary for the world to be best need not be necessary for *our* well-being.

We might be tempted to conclude that, for us, the best of all possible worlds is a blight rather than a blessing. To suppress the temptation, Leibniz assures us that the injustices we suffer in this life will be compensated in an afterlife. He accepts Augustine's suggestion that evil is a privation, and therefore not actual. The actual world is "the most perfect morally," and this moral perfection makes this world best for us.

VII
SECULAR NOTHING

19
Schopenhauer
Absence of Meaning

Gottfried Leibniz's optimism filled a much-needed gap in metaphysics. Arthur Schopenhauer (1788–1860) inverts Leibniz's doctrines to restore the much-needed gap.

Teenage Antitheodicy

In 1737 the Jesuit journal *Memoires des Trevoux* introduced *optimisme* to describe Leibniz's doctrine in the *Theodicy* that the actual world is the best of all possible worlds.

Inspired by Voltaire's parody of Leibniz's *Theodicy* in *Candide*, Schopenhauer turned optimism upside down (fig. 19.1).

Ours is the worst of all possible worlds. Suffering is caused by disorder. The actual world manages to be just orderly enough to achieve more misery than any other possible world.

Just as Leibniz tried to explain away apparent evil as necessary for hidden goodness, Schopenhauer explained apparent good as necessary for hidden evil. Sexual pleasure exists only to draw more sufferers into existence. A drop of beauty provides the contrast for a vista of ugliness. Sprinkles of good fortune are needed to prolong and deepen despair.

Goethe's *Faust* may have Satanized Schopenhauer's early inversion of Leibniz. With "the spirit of perpetual negation" in charge, the meaning of life would be suffering.

With a few more diabolic adjustments, the rest of Leibniz's heaven is reprogrammed into Schopenhauer's hell. The principle of sufficient reason now ensures that nothing goes to waste in the maximization

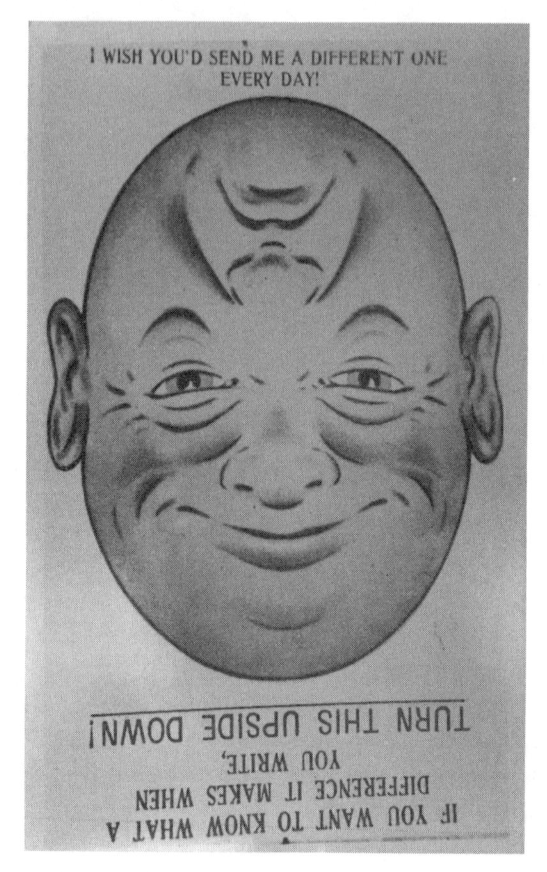

Figure 19.1 Happy Leibniz (invert to see Schopenhauer)

of misery. Panpsychism now makes the whip suffer along with the whipped.

Optimism is itself a tool of suffering. Optimists divert attention from evils that can be eliminated and dangers that can be reduced. Optimists exaggerate what is good, excuse what is bad, and gamble what little we have on ventures that leave us with even less. Expectations climb to maximize the inevitable fall. Schopenhauer renders a service to humanity by exposing the hidden cruelty of optimism.

Whereas Leibniz follows Augustine in downplaying evil as an absence of good, Schopenhauer downplays good as an absence of evil: "I know of no greater absurdity than that propounded by most systems of philosophy in declaring evil to be negative in its character. Evil is just what is positive; it makes its own existence felt" (1851, 5). Pain signals tissue damage and thereby commands primary attention from the sufferer. Pleasure is mere absence of pain, as when a toothache subsides. Health is the absence of disease. Peace is the absence of war. Rationality is the absence of irrationality.

Rightness is nothing but the absence of wrong. Whereas Leibniz is an activist racing around Europe on diplomatic missions, Schopenhauer minds his own business. Omissions are morally safer than actions. Nonvoters are told they have no right to complain when elected official misgovern. But Schopenhauer would counter that if anyone has lost the right to complain, it is the voters. They *did* something bad. The nonvoters merely let something bad happen.

Schopenhauer would rather be ruled by a lion than by a fellow porcupine:

> On a cold winter's day, a group of porcupines huddled together to stay warm and keep from freezing. But soon they felt one another's quills and moved apart. When the need for warmth brought them closer together again, their quills again forced them apart. They were driven back and forth at the mercy of their discomforts until they found the distance from one another that provided both a maximum of warmth and a minimum of pain. In human beings, the emptiness and monotony of the isolated self produces a need for society. This brings people together, but their many offensive qualities and intolerable faults drive them apart again. The optimum distance that they finally find that permits them to coexist is embodied in politeness and good manners. Because of this distance between us, we can only partially satisfy our need for warmth, but at the same time, we are spared the stab of one another's quills. (1851, 651–652)

Schopenhauer took pride in his ignorance of politics. His only foray into the public sphere is literary. Instead of forcing anyone to do anything, he exposes the terrible truth for the few sober enough to heed it.

Leibniz "proves" theism by defining God as 'the best conceivable being'. Schopenhauer feels that the pessimist could equally well "prove" the Devil exists with the definition 'the worst conceivable being'. "Night is falling: at dusk, you must have good eyesight to be able to tell the Good Lord from the Devil" (Jean-Paul Sartre *The Devil and the Good Lord Act* 10, sc. 2). Leibniz's ontological argument rests on the principle that existence is better than nonexistence. Schopenhauer counters that existence is worse than nonexistence.

Since being is always contingent, each entity is a gratuitous evil. According to Schopenhauer, this is the insight behind the Christian doctrine of original sin. In Jean-Paul Sartre's 1938 *Nausea*, this epiphany is generalized beyond people to chestnut trees and inanimate objects. The main character experiences existence as *de trop*: too much, excessive, overdone. Reality is as overstuffed as Noah's Ark.

The universe, as the sum total of all there is, is the biggest mistake. There are no necessary evils. All evils are gratuitous.

Leibniz extols permanence (continuing the Egyptian tradition, rationalized by Plato). Schopenhauer champions the ephemeral. After all, if existence is bad, brevity is a virtue. Better to be a mayfly than a dwarf pygmy goby (which lives two months rather than a day). Better to be a dwarf pygmy goby than a Greenland Shark (which can last four centuries). Never to have been is best of all.

Since Schopenhauer is inverting Leibniz's system, these opposite thinkers are wedded to the same problematic presuppositions. If existence is better than nonexistence, we must compare what is to what is not. But what is not, is not available for comparison. Both the optimist and the pessimist are stuck with an apparently impossible comparison between being and nonbeing.

The Business of Life

The mature Schopenhauer was too much of a Shakespearean to endow our lives with cosmic significance.

> Life is but a walking shadow, a poor player,
> That struts and frets his hour upon the stage.
> And then is heard no more. It is a tale
> Told by an idiot, full of sound and fury,
> Signifying nothing. (*Macbeth*, act 5, scene 5, ll. 17–28)

The proper negation of theism is drab atheism, not colorful Satanism.

For atheism to mushroom out of Leibniz's corpus, Schopenhauer needed to exorcize the Devil. The method emerged from the great curse of Schopenhauer's formative years—economics.

"The Protestant pastor is the grandfather of German philosophy" (*The Anti-Christ*, sec. 10). So said the son of one Protestant pastor, Friedrich Nietzsche (recalling Kant, Fichte, Schelling, and Hegel). The exception was the philosopher who most influenced Nietzsche: Schopenhauer.

Arthur Schopenhauer's father Heinrich was a successful entrepreneur. At 38, Heinrich married the 18-year-old Johanna Trosiener. She would later become the best-known female author in Germany, the first not to employ a pseudonym. The couple reared their son and daughter in a well-rounded bourgeois setting. Religion played only a conventional role. Mr. and Mrs. Schopenhauer instilled appreciation of art, literature, and science.

There was a meaning to Arthur's life. As a meticulously planned child, he was conceived with the intention that he take over the family business. His very name had been chosen for its cosmopolitan, mercantile virtues; 'Arthur' is pronounced and written the same way in German, French, and English.

Arthur stifled his passion for literature. He boarded at a British business school. The institution reinforced commercial skills with religious ritual. Young Schopenhauer also learned the practical details of trade by apprenticing with leading merchants. His father's letters stress the importance of appearance: *Arthur, don't slouch! Arthur, practice your penmanship!*

Arthur's reverence for his domineering father morphed into ambivalent admiration for the traits Charles Dickens mocks with his character Ebenezer Scrooge. The merchant openly and honestly seeks what others only surreptitiously pursue: profit. The merchant meets his obligations and insists that others meet theirs. He does not suffer fools. He sees through parasites in their sanctimonious guises (such as employees who think fairness requires them to receive something for nothing, such as paid holidays). The merchant is not fooled by fictions. The ledger book trumps all other forms of literature. Everything reduces to the bottom line.

On April 20, 1805, Heinrich Schopenhauer fell from a tall building. Staggered by what he and his mother secretly believed to be a suicide, the seventeen-year-old Arthur reassessed his life . . . and Life. Arthur resented his socialite mother's failure to nurse her husband through his mental and physical illness. Arthur also resented her interest in keeping her son in business so that she and her daughter could continue their lavish lifestyle. For two years Arthur complied postmortem with his father's career wish. The dead hand of the past was strangling his future. Arthur finally broke his commitment to the family business—selling out so that he could retire as an independent scholar.

False Desire

The decision to go out of business is akin to the decision to kill yourself. The merchant ranks potential lifetimes of a firm. Rarely is the longest lifetime the most profitable lifetime.

Economists model the apprentice as an agent who rationally calculates which firm to join. Schopenhauer favors a more anthropological perspective. Apprentices are often too immature and ill-informed to make an enlightened choice. Family members, tradition, and circumstances induct them into the firm. Apprentices lose more by forfeit than by active contest.

Economics is a science of means, not ends. Only instrumental desires are assessed—and always by the criteria of efficiency. Schopenhauer, however, had become persuaded that there are false desires. Lunatic impulses, weakness of will, addiction, and brainwashing show there can be desires that ought not to be acted upon. "Wealth is like seawater; the more we drink, the thirstier we become" (1851, vol. 1, 347). Desire can be displaced: "Money is human happiness in the abstract; he, then, who is no longer capable of enjoying human happiness in the concrete devotes himself entirely to money" (1912 vol. 2, § 320). False desires also arise in animals. Parasites reprogram their host's psychology to suit their own interests. Advertisers beguile customers into wanting what is unnecessary or even injurious.

We generally assume that an apprentice can transfer to a firm that is free of pathological desires. Schopenhauer challenges this assumption. The very logic of desire ensures that "life is a business that cannot cover its cost." This is even truer in transactions between different parties. Schopenhauer continues:

> Pleasure is never as pleasant as we expected it to be and pain is always more painful. The pain in the world always outweighs the pleasure. If you don't believe it, compare the respective feelings of two animals, one of which is eating the other. (1851, 5–6)

Our desires form a queue: water, food, sex, sleep . . . Unrequited desire is distressing. The relief afforded by its satisfaction is of little moment. Satisfying one desire merely exposes its impatient successor. The queue is disorderly. Wish elbows out want. Whim cuts ahead of need.

There are occasions in which no new desire emerges. If this absence were experienced as a recess, the server might cultivate by design what initially happened by accident. Accordingly, the server is built to experience this appetitive void as unpleasant—as boredom.

The queue of desire is a kludge—an inelegant but adequate compatchment, cobbled together with what was available. When a castle is retrofitted into a prison, the inmates may discover recesses, affordances, and even the means of escape. Although boredom mortars up a promising egress, there are some openings offered by art and scientific contemplation. For short breaks, we experience objects without a view to their practical utility. Geniuses resemble madmen in their ability to sustain this detachment for extended periods.

Saints and fakirs mitigate misery through the suppression of desire. They uproot the source of suffering—desire itself. Schopenhauer finds this the most admirable solution (although he only adopted the milder strategies discussed in the previous paragraph).

Economists prize impossibility theorems in which a set of desiderata is shown to be impossible to fulfill. Schopenhauer's pessimism is based on an impossibility theorem: *every* set of desires is pathological. An impartial inspection of the ledgers of life discloses a structural deficit.

Impartial self-appraisers do not multiply. Indeed, realists subtract themselves from the population. The remainder cook the books. Schopenhauer uses literary techniques to spring a surprise audit on the ledger of life.

Schopenhauer is ambivalent about reason. On the one hand, he is an optimist about the power of truth to overcome error. To correct false conclusions, Schopenhauer ferrets out fallacies. As a novice lecturer, Schopenhauer scheduled himself against the most popular lecturer, Georg Hegel, whom Schopenhauer dismissed as a charlatan. With an advantage of a convenient side-by-side comparison, students would recognize Schopenhauer as more logical. Schopenhauer lectured to nearly empty rooms. Later, Schopenhauer predicted that Buddhism would prevail over other religions because of its favorable ratio of truth to myth. Lastly, Schopenhauer predicted that his own philosophy would become popular in the long run. His otherwise admiring commentator Søren Kierkegaard noted the irony; Schopenhauer fails to see that pessimism precludes belief in progress and just deserts.

Despite this de facto respect for the power of truth, Schopenhauer denied that logic has any practical value. Validity is conditional; the conclusion is true if the premises are true. "Reason is feminine in nature; it only gives after it has received. Of itself it has nothing but the empty forms of its operation" (Schopenhauer 1958, 2:50) Our brains run on blood rather than syllogisms. Our actual inferences are like footsteps: automatic, contextual, and directed toward biological goals. Consequently logic, despite being perfected by Aristotle long ago, has only theoretical significance.

Schopenhauer is influenced by German Romantic poetry (Millán 2017). People do not need logic to reason, just as musicians do not need notated frets to play the guitar. Logic merely summarizes intuitive assessments of valid inference. Whenever there is any conflict between theory and intuition, logicians revise theory rather than intuition. Proofs are like mousetraps. They lure step by step, with an illusory prize. When the trap is sprung, we are pinned into our captor's stunning position—but without any enlightenment.

Our refusal to rely on logic is healthy. For logic distorts reality. The law of excluded middle requires continuous phenomena to be shoehorned into discrete categories. The brain was no more designed to do philosophy than the stomach; both are servants of the gonads. Accordingly, Schopenhauer's expectations for academia are low. Young sea-squirts swim about feverishly seeking a place of permanent rest. Upon success, they absorb their brains—like tenured professors.

The Invisible Hand of Suffering

In *The Frogs*, Aristophanes laments the decline of Athens from a golden age. A pair of concerned citizens descends to Hades to resurrect the old leadership. Bad politicians have driven all of the good statesmen out in the same way as bad money drives out good money.

How does the bad drive out the good? Queen Elizabeth inquired after she ordered new coins to be minted with less gold. The English financier Thomas Gresham (1519–1579) quieted the royal curiosity: coins composed of the cheaper metal were used for payment; coins made of the more valuable metal were hoarded.

No financial devil is needed to explain how overgrazing ruins public pastures. Each shepherd merely acts from self-interest, not malice. When a shepherd understands why the pasture is deteriorating, his reaction is to graze his sheep *more* intensely—hastening before all the grass is consumed. All public goods tend to be eaten away. A sheriff might police the pasture. But law enforcement is just another public good having the same tendency toward decline (by those who want the service of the sheriff but do not want to contribute to his salary).

Good politicians suffer asymmetrically from betrayal, smears, and the effects of bribes and false promises made to the public. Corrupt politicians displace the virtuous politicians.

The prevalence of incompetence in firms is explained by the tendency of employees to be promoted until they manifest incompetence. Since incompetent employees are difficult to fire and cannot find alternative employment at the same level, they remain at post, performing poorly until retirement.

Thomas Malthus predicted that human populations would expand until resources ran out. Population growth increases geometrically. Food increases only arithmetically. So unchecked population growth will lead to starvation and conflict.

Charles Darwin generalized Malthusian reasoning. Organisms overbreed, crowding each other to the precipice of nonexistence.

Schopenhauer anticipates Darwin by envisaging a similar threshold. Organisms compete to the limits of their endurance. The unintended effect of the Will to Life is the maximization of misery. It is as if the whole universe is fined tuned for discord and dissonance.

As Sophocles observed, the delights and satisfactions of life are co-opted to promote long-term suffering: "Man's pleasure is the spring of sorrow" (*Antigone*, l. 621). Even humor is merely a means of increasing our ability to endure suffering. Inscription on a physician's tombstone: "Here lies he like a hero, and those he has slain lie around him" (Schopenhauer 1819, 2:273) Logically, we ought to be disheartened by the physician's incompetence. But we instead find a perverse enjoyment of the suffering. Gallows humor is admired all the more when voiced by the victim. As the disgraced, destitute Oscar Wilde lay dying in a shabby hotel room, his last words were: "Either that wallpaper goes or I do." Or so say fans in preference to him parting with a mumbled Catholic prayer.

Immanuel Kant, who Schopenhauer revered, comments on the mechanics of jokes: we experience comic relief when "a tense expectation is transformed into nothing." Kant's specimen underscores the hydraulic nature of humor. "An Indian at the table of an Englishman in Surat, seeing a bottle of ale being opened and all the beer, transformed into foam, spill out, displayed his great amazement with many exclamations, and in reply to the Englishman's question 'What is so amazing here?', answered, 'I'm not amazed that it's coming out, but by how you got it all in' " ([1790] 1911, First Part, sec. 54).

Art provides another Kantian eye-opener. Instead of perceiving the object in terms of its potential service, you adopt an impartial perspective. You are temporarily absented from means-ends striving. In the case of instrumental music, the nonrepresentational nature of the experience leads Schopenhauer to suggest that one directly experiences reality, that is, the Will to Life.

The inferno of life is more rapacious than fire. Whereas fire burns itself out, life feeds on life. The self-consumptive aspect of life is occasionally exemplified within a single organism.

> The bulldog-ant of Australia affords us the most extraordinary example of this kind; for if it is cut in two, a battle begins between the head and the tail. The head seizes the tail in its teeth, and the tail defends itself bravely by stinging the head: the battle may last for half an hour, until they die or are dragged away by other ants. (1819, 1:147)

The tail of the bulldog ant will reflexively sting anything. It has no free will. The head is no freer than the tail. The fact that the head *would* have done differently is of no more significance than the fact that the tail *would* have not stung the head had there been a different stimulus.

Your sexual organs incarnate the drive toward reproducing life. Your digestive organs incarnate the drive toward sustaining your life. To sustain your life, you must destroy other life. In 1911 Ambrose Bierce's *The Devil's Dictionary* gave a Schopenhauerian definition of 'edible': "Good to eat, and wholesome to digest, as a worm to a toad, a toad to a snake, a snake to a pig, a pig to a man, and a man to a worm."

The normal purpose of an invisible-hand explanation of a bad outcome is to exonerate the participants; they are merely self-interested, not malicious. However, Schopenhauer blames us for frustrating the wills of others. In principle, we could avoid wrongdoing by not acting. To live is to kill. Consequently, the cycle of destruction is rough justice. Schopenhauer views this as a vindication of the Christian doctrine of original sin. We deserve to suffer.

Nihilating Dichotomies

Suffering requires a sufferer. There must be a contrast: self/other, agent/patient, self/environment, and so on. Erasing either side of the dichotomy eases suffering.

Egoism is the overwhelming force in human affairs. There is relief in the temporary selfless standpoint of ethics.

In detached observation, the scientist escapes conflict with his environment by become engrossed in it. Ditto for the artist who approaches the art object as a disinterested spectator. Accordingly, Schopenhauer warns against depicting food because it may stimulate hunger. Nudes spark lust. They teeter toward pornography. The safest object of contemplation is instrumental music.

The sublime involves a reciprocal annihilation. When a mountaineer views an approaching storm, sheer scale reduces him to insignificance. Given the idealist equation "To be is to be perceived," the mountaineer will simultaneously be aware of how the storm depends on him. Without his representation, the storm is nothing.

Absence of Self

Leibniz defended freedom by portraying each individual as following his own program. Instead of being caused by external events, each individual is synchronized with every other individual in the orchestra. Each of us is self-caused in the sense that we act out of our inner nature. We each have destiny that harmonizes with destinies of every other thing. We fit in with the universe without being subjugated by it.

According to Schopenhauer, your life is abstracted out of the great mass of Life. Sculptors imagine a figure in a block of stone and then chisel everything else away). Your life is mentally carved from Life. But no sculptor actually separates you.

There are statues depicting a muscular sculptor carving himself into existence from a stone block (fig. 19.2). This statue of a self-made man is a parallel of Leibniz's God who draws himself into existence. The two depictions are equally incoherent. A creator must preexist the creation. Nothing can precede itself.

Since there is no self, there is no possibility of self-destruction. As figure becomes ground, you merely become a less I-catching part of Life. Schopenhauer interprets paranormal phenomena as echoes from the mental lives of the dead. They linger on in a hellish cacophony,

Figure 19.2 Bobbie Carlyle's *Self-Made Man* at the Lay Center, Saint Louis University

becoming ever less distinct. You do not end in silence. You continue as white noise.

Instead of being a genuine denial of life, suicide is an incoherent affirmation of it. The pitiable sufferer attempts to flee from being to non-being. But there is no destination for this flight. Schopenhauer grants a depersonalized form of immortality—of Life after your "life."

Human Being Is the Worst Being

Homo sapiens has a cognitive niche, exploiting the advantages that accrue from the prediction and control of phenomena. Whereas other organisms live in the present, people are future minded. Since the past is a guide to the future, human beings expand their temporal horizon in both directions.

Schopenhauer compares the human present to a mobile cloud that dims one's current location, intensifying all other times by contrast. The cloud moves with the direction of time—toward death.

What is past, no longer exists. Every something will become a nothing. Our moments begin at the top of a bottomless hourglass. Each moment drops from the Future through the bottleneck of the Present into the abyss of the Past.

The Need for Metaphysics

Optimism is the attitude most conducive to reproduction. People innately believe that the purpose of life is happiness. When suffering violates this birthright, people ask why.

One explanation, familiar from childhood, is that the torment is a *deserved* response to some infraction. Salvation requires that we identify who has been offended, understand the nature of the offense, and atone.

However, all efforts to find the hidden punishers fail. Nobody is there. Metaphysics commences when we evade these empirical refutations. Punishers become intangible, inaudible, invisible . . . As we attribute more powers to these hidden agents, fewer of them are needed to explain our misfortunes. This permits vertical integration of our supplications. Inferior gods are rendered redundant by superior gods. Eventually, one all-powerful, all-knowing god emerges from the crowd, capable of any feat.

Optimistic versus Pessimistic Religions

Optimistic religions, such as paganism, pander to our biases by reassuring us that life is good. Pagans claim our earthly lot will be improved by placating supernatural governors.

There are no such beings. In consequence, the monotheism of the Jews is closer to the truth than polytheism. ("Only one god off!" runs the atheist's compliment). The price of this incomplete ontological progression is religious intolerance.

Judaism retains the optimism of the pagans. Their God made the world by caprice. God applauds his own work.

Christianity triumphed over Judaism by facing up to the truth that life is terrible. Christian world-rejecters expose optimistic biases that discourage distractions.

In addition to the gain in impartiality, the Christians correctly diagnose the underlying source of suffering. Instead of being an accident, our suffering is systematic, predictable, even lawful and just. We have infinite desires but only finite means. Earthly frustration is inevitable.

Impressed, Schopenhauer speculates that the Christian analysis of suffering is plagiarized from the yet deeper diagnosis achieved by the superior religions of India (perhaps mediated through Egyptian priests). Brahmanism and Buddhism delve into the irreality of the self and objects. Since the Indians are closer to truth, Schopenhauer is not surprised by the failures of missionaries in India. Why should the Indians convert to the more primitive religion of Christianity? Instead of the West converting the East, the East will convert the West. The Oriental renaissance will be as profound as the Renaissance based on the rediscovery of ancient Greek wisdom.

Schopenhauer acknowledges that the Muslims have made conversions in India—at the point of a sword. This is regress because Islam is even more backward than Judaism (which at least does not proselytize to non-Jews):

> Temples and churches, pagodas and mosques, in all countries and ages, in their splendor and spaciousness, testify to man's need for metaphysics, a need strong and ineradicable, which follows close on the physical. The man of a satirical frame of mind could of course add that this need for metaphysics is a modest fellow content with meager fare. Sometimes it lets itself be satisfied with clumsy fables and absurd fairy-tales. If only they are imprinted early enough, they are for man adequate explanations of his existence and supports for his morality. Consider the Koran, for example; this wretched book was sufficient to start a world-religion, to satisfy the metaphysical

> need for countless millions for twelve hundred years, to become the basis of their morality and of a remarkable contempt for death, and also to inspire them to bloody wars and the most extensive conquests. In this book we find the saddest and poorest form of theism. Much may be lost in translation, but I have not been able to discover in it one single idea of value. Such things show that the capacity for metaphysics does not go hand in hand with the need for it. (1819, 2:162)

The Complicity of Rationalism

Rationalism is an attempt to rescue the comforts of religion. To counter the empirical evidence, the rationalist reminds us of occasions on which reason overrides the senses. Our eyes tell us that the earth is flat and stationary. Reason reveals it to be round and moving.

Massive suffering demonstrates a posteriori that there is no all-powerful, all-knowing, all-good God. But the rationalist's ontological argument for God counterdemonstrates a priori that his existence is true by definition! Accordingly, Leibniz infers that evil is a fictitious force. On a merry-go-around, you feel there is a force pushing you away from the center. But centrifugal "force" is an illusion generated by your limited perspective. An observer standing outside your frame of reference will only detect the centripetal force needed to keep you from traveling in a straight line.

According to Leibniz, we rely on experience only because we have insufficient computing power. Experience is a mere shadow of reason—a chain of associations that links us with animals. For God, every statement is an analytic truth. Instead of being a positive source of knowledge, experience is a humiliating reflection of our limits.

Containing Reason

Whereas Leibniz used the principle of sufficient reason to expand the realm of the explicable, Schopenhauer wrote his dissertation *The Fourfold Root of the Principle of Sufficient Reason* to demonstrate the limited scope of reason. He agrees with Leibniz that there is a reason for each thing. However, Schopenhauer insists that each explanation must start with a particular thing. Thus reason cannot answer "Why is there something rather than nothing?" Nor can anything else. The existence of the Will to Life is a brute fact.

The category of the thing in question dictates which kind of explanation is feasible. All explanation proceeds by revealing necessary connections. Like only connects to like. Material things must be grounded to every other thing in space and time but can only link with other material things. Abstract concepts can only connect other abstract concepts (as in definitional explanations). The same segregation applies for mathematical constructions and psychologically motivating forces. There are only contingent connections *between* categories. Just as offspring are only produced from parents of the same species, conclusions are only produced from premises within the same category of things. The German idealists of his era (Fichte, Schelling, and Hegel) violate these boundaries.

Double-Aspectism

The correct relationship between the thing-in-itself and our sensations is akin to the relationship between the convex and concave surfaces of a bowl. Each side is an aspect of the same thing. The same goes for your body. You are aware of it from the outside as a physical object. But you are also aware of it from the inside, by how it feels—as will. When you cup your hands, you are not aware of first forming the intention and then the hands coming together. Rather, you feel the intention and the movement simultaneously as two aspects of the same thing. If you neglect the subjective side, the result is materialism, "the philosophy of the subject that forgets to take account of himself" (Schopenhauer 1958, 2:13).

Your direct access to other people is as physical objects. But from your own case you realize their bodies have a subjective aspect.

Most people project psychology only partway down the chain of being, stopping at mammals, or fish, or perhaps insects. Since life is continuous, these demarcations are arbitrary. Schopenhauer extends psychology universally:

> Spinoza says that if a stone which has been projected through the air, had consciousness; it would believe that it was moving of its own free will. I add this only, that the stone would be right. The impulse given it is for the stone what the motive is for me, and what in the case of the stone appears as cohesion, gravitation, rigidity, is in its inner nature the same as that which I recognize in myself as will, and what the stone also, if knowledge were given to it, would recognize as will. (1819, 1:126)

As a panpsychist, Schopenhauer has no problem explaining how the mind and body interact. What was never sundered has no need for reunification. Nor does the panpsychist have any problem knowing there are other minds. Physicality without mentality is impossible. *Necessarily*, there are other minds.

Admittedly, the psychological side, the Will, is not directly accessible—with the single exception of one's own body. Each of us can universalize this point. Each of us is the center of an all-encompassing, infinite sphere. Each of us is a microcosm.

Empiricism and Science

Rationalists are optimists. They regard reality as governed by reason and reason as responsive to desert. So what is, is good—and getting better.

Empiricists are pessimists. 'Everything we know about the world is based on experience' commits them to debunking transempirical consolations. Any source of value beyond the satisfaction of desire is forbidden. Since empiricists accept the value neutrality of nature and also regard it as indifferent to desire, they predict much pointless suffering.

Historically, rationalists precede empiricists. Thus empiricists enter the dialogue reactively, responding to a conversation that rationalists had previously monopolized. Since empiricist reject synthetic a priori propositions, they strip many of the traditional consolations fabricated by rationalists.

A few consolations remain. Albert Einstein had a photograph of Schopenhauer in his 1918 Berlin study:

> I believe with Schopenhauer that one of the strongest motives that leads men to art and science is escape from everyday life with its painful crudity and hopeless dreariness, from the fetters of one's own ever shifting desires. A finely tempered nature longs to escape from personal life into the world of objective perception and thought; this desire may be compared with the townsman's irresistible longing to escape from his noisy, cramped surroundings into the silence of high mountains, where the eye ranges freely through the still, pure air and fondly traces out the restful contours apparently built for eternity. (Einstein [1918] 2002)

Einstein gained patience from Schopenhauer's determinism.

Schopenhauer would have been dismayed by Einstein's later lapses into rationalism. Einstein grew sympathetic to Spinoza's identification of God with nature. Pantheism is cockeyed: "For from a first and impartial view, it will never occur to anyone to regard this world as a God. It must obviously be an ill-advised God who could think of no better amusement than to transform himself into a world like the present one" (1851, 41)

Empiricism is more congenial to science and is less able to evade common-sense challenges to religion. It also elaborates on common-sense evidence that the universe is indifferent to us. Astronomers have shown that the earth is not in any favored position in the solar system or the galaxy.

Since science uses discrete concepts to categorize a continuous reality, it is inherently distortive. Nevertheless, science can still support philosophical theses. Schopenhauer collected examples of scientists vindicating tenets of his system.

In 1852 Lord Kelvin introduced the notion of heat death. The law of thermodynamics suggests the ultimate total destruction of all we care about. As energy runs down the hierarchy of levels, everything trends toward an undifferentiated soup of particles. Since 'nothing' can be used to mark absence of differentiation, the trend is toward a universal nothingness, "heat death," in which there is no significant differentiation.

Just 10 months before Schopenhauer's death, Charles Darwin published *The Origin of Species*. This book put many of Schopenhauer's theses on a professional scientific footing. The blindness of the Will to Life is reflected in Darwin's principle that natural selection has no foresight; nature cannot take one step back to make two steps forward. However, Schopenhauer misses the principle of natural selection. He has a Platonic conception of species as prototypes. This makes individual differences between members of the same species irrelevant.

Darwin is more optimistic than Schopenhauer. Darwin thinks species tend to progress. He allows inheritance of some acquired characteristics. So the benefits of hard work, such as a giraffe stretching its neck to reach taller leaves, could be passed to future generations. Concluding allusions to a Creator become more pronounced in subsequent editions of the *Origin of Species*.

Schopenhauer thought personal experiences confirm his magisterial pessimism. When friends wrote him of their troubles, Schopenhauer consoled them with the value of their experiences as confirmation of his philosophy: "Life can be compared to a piece of embroidered material of which, everyone in the first half of his time, comes to see the top side, but in the second half, the reverse side. The latter is not so beautiful, but is more instructive because it enables one to see how the threads are connected together" ([1890] 2007, 102).

20

Bergson

The Evolution of Absence

Charles Darwin's (1809–1882) *On the Origin of Species* was initially regarded as a dispiriting confirmation of Schopenhauer's atheistic pessimism. Herbert Spencer (1820–1903) spotted a silver lining in this dark cloud. Life gets better because there is "survival of the fittest." This improvement compensates for the suffering. If we let nature take her course, the human race will incrementally achieve its destiny. If we intervene in the right direction, progress accelerates. Any paddling by social reformers should therefore be downstream. Paddling upstream, against the evolutionary current, would increase misery in the long term.

French intellectuals embraced this progressivist interpretation of evolutionary theory. Henri Bergson (1859–1941) tried to reconcile Catholics to Darwin by suggesting that God operates through the invisible hand of natural selection. Only mystics feel the touch of the Hand directly. The rest of us only sense the divine secondhand. "Religion is to mysticism what popularization is to science" ([1932] 1977, 204). Voilà! Schopenhauer's *Wille zum Leben* was translated into Henri Bergson's *élan vital*.

Unmollified, the Catholic Church placed Bergson's *Time and Freewill*, *Matter and Memory*, and *Creative Evolution* on the Index of prohibited books (Decree of June 1, 1914). Bergson is a dynamic pantheist. God bubbles up endlessly from the process of his own creation. Bergson continued to court the church by hints that he might convert to Catholicism.

Bergson's Creative Impulse underscores the positive nature of reality. Leucippus had pioneered negative philosophy by postulating the void. Christian resistance had been overcome by Isaac Newton. The success of physics gave rise to a mechanical philosophy of discrete changes from sharply defined atoms. Cold was defined as the absence of heat. Darkness as the absence of light. Silence as the absence of sound. After Newton's divinization of space ebbed, only space's austerity remained: cold, dark, silent, in no way dependent on objects in space. Emptiness was the default state. Across the border in Germany, idealism converged with atomistic materialism by treating being and nonbeing as equally real. The tension between them gives rise to becoming.

Perceiving Nothing?

To perceive is to perceive something. That something must be a cause of the perceptual experience. So we cannot perceive nothing. That would violate both the intentionality of perception (to perceive is to perceive something) and the causal requirement (one can perceive X only if X is a cause of its own perception). Yet we see shadows (absences of light), hear silence (absence of sound), and feel holes (absences of matter).

The leading German philosopher Georg Hegel (1770–1831) tried to resolve our predicament with metaphysical egalitarianism. Recalling the ancient Greek atomist Democritus, Hegel ranks being and nonbeing as equally real. There is something to see in a lightless cave: nothing. There is something heard in a soundless chamber; nothing.

The French, under the leadership of Bergson, took it as axiomatic that reality is fundamentally positive. Perceptions of absences had to be explained away.

Absences as Spectacles

When the *Mona Lisa* was stolen on August 21, 1911, its absence passed unnoticed by officials at the Louvre. On August 22, an artist who had made a special trip to copy the *Mona Lisa* complained to a guard. The Louvre was searched, Searched, SEARCHED. Nothing! The police closed the Louvre for a week to conduct an investigation of the staff. When it reopened, a long queue formed to ogle the absence of *Mona Lisa*. More Parisians came to her absence than her presence (fig. 20.1). The daily morning newspaper *Le Figaro*, founded in 1826 and still publishing, described the spectacle as "an enormous, horrific, gaping void." According to its reporter, "The crowds didn't look at the other paintings. They contemplated at length the dusty space where the divine Mona Lisa had smiled the week before."

Figure 20.1 Vacant wall in the Louvre's Salon Carré after the *Mona Lisa* was stolen in 1911

The "wall of shame" was kept vacant for weeks to accommodate demand. This permitted participation by foreign tourists, such as Franz Kafka and his friend Max Brod. The pair continued their pilgrimage the following evening by attending the cinema to see a parody: *Nick Winter and the Theft of the Mona Lisa*.

Postcards were printed so that the pious could pray to Saint Anthony, the patron saint for things gone missing (fig. 20.2). When the postcard is held to candlelight, the heat makes the *Mona Lisa* appear.

Figure 20.2 Saint Anthony postcard

Catholics pray to St. Anthony to recover keys, wallets, and poodles. But they cannot pray to recover their lost faith. If you believe that God does not exist, then there is no one to pray to. 'Atheists worship nothing' is properly read as characterizing these nonbelievers as nonworshipers rather than as worshipers of nothing. We might try the same response to those who claim that the curious Parisians came to perceive nothing. Under this nonperceptual hypothesis, the Parisians came to the Louvre to not perceive the *Mona Lisa*.

Anomaly: why would the Parisians travel to the Louvre for the purpose of not seeing? Atheists do not go to church to not pray!

Unlike the atheists, the Parisians intended to have a certain kind of visual experience that was coupled to a specific site. The blind stayed home.

Yet the curious Parisians seemed to concede that there was nothing to look at. If the curious Parisians had seen the *Mona Lisa* at the Louvre, they would have been surprised by the absence of the absence of the *Mona Lisa*.

Objectless Seeing?

Were the curious Parisians trying to enter a perceptual relationship that does not require anything to look at? Although 'see' is generally read as a transitive verb, some kinds of visual experiences do not require anything *external* to be seen.

Leonardo da Vinci asks, "Why does the eye see a thing more clearly in dreams than the imagination awake?" The implied answer is that some visual experiences proceed directly. The dreamer does not need to cope with distance, fog, or myopic eyes.

However, the curious Parisians were interested in seeing something public. Dream seeing is private. The absence of the *Mona Lisa* was in the common world, more precisely, in the Salon Carré.

Visual imagery can be presented as being in public space and can involve processes such as recognition. Stare at the photographic negative in figure. 20.3.

Figure 20.3 Negative inverse

Redirect your gaze to a blank page. You will experience a negative afterimage. You may recognize the image as that of the *Mona Lisa*. Although projected into a public space, the image is private.

Negative afterimages involve fatiguing the eye. There are social sorts of visualizing that involve deeper alterations of the visual system. But the Parisians were not on a vision quest. They were not cultivating a collective hallucination. Nor were they attempting to perceive spiritual beings as when they attended séances. The Parisians went to the Louvre sober and fit. They were engaged in the observation of a well-lit, public phenomenon.

The Parisians regarded the absence of the *Mona Lisa* as something that could be photographed. And they actually photographed it!

Romantic Defeatism?

People who go blind continue to dream visually. While awake, many visualize the scene before them. They differ from congenitally blind people who miss a developmental opportunity for the visual system to develop. In this sense, completely blind people vary in their ability to see. All fail to see but some more closely approximate seeing than others. Those who visited the Louvre came closer to seeing the *Mona Lisa* than those who stayed home. Perhaps they merely wished to approximate seeing the *Mona Lisa*—without actually seeing.

In Frank Capra's 1946 movie *It's a Wonderful Life*, a girl, Mary Hatch, whispers into the deaf ear of the adventurous boy George Bailey: "Is this the ear you can't hear on? George Bailey—I'll love you till the day I die." Mary expresses her affection by coming close to letting him hear her declaration of love. The French museum visitors may have been engaged in similar symbolism. The crowd was protesting the fact that the *Mona Lisa* could no longer be viewed at the Louvre.

But many of the curious Parisians regarded their trip to the Louvre as a success rather than a noble failure. They jostled to get a good look at the absence of the *Mona Lisa*. Short Parisians at the far end of the queue complained that they could not see anything. All had waited in line for the opportunity to view the absence of the *Mona Lisa*. When their turn came, they adopted the optimal viewing angle (relative to the eye level of the observer and distance to the painting).

In sum, the curious Parisians put us in a dilemma. We cannot find a perceptual object for their visit to Louvre. Yet we cannot charitably interpret them as seeking to not see the *Mona Lisa*.

Bergson's Metacognitivism

Just three years prior to the theft of the *Mona Lisa*, the most famous French philosopher of the era, Henri Bergson, claimed to have refuted the possibility of perceiving absences. In the final chapter of *Creative Evolution* Henri Bergson writes:

> When I say, "This table is black," I am speaking of the table; I have seen it black, and my judgment expresses what I have seen. But if I say, "This table is not white," I surely do not express something I have perceived, for I have seen black, and not an absence of white. It is therefore, at bottom, not on the table itself that I bring this judgment to bear, but rather on the judgment that would declare the table white. I judge a judgment and not the table. (1911, 303–4)

Logically, negative "perceptual" reports are a form of metacognition that gets confused as object-level perception.

Negation is never "a complete act of the mind" (Bergson 1911, 287). Instead of recording just how things are, negative judgments detour into the realm of possibility.

The idealist logicians had already refined Kant's doctrine that negative judgments are higher-order judgments: "The task peculiar to negative judgments is to reject error" ([1781] 1965, A709/B737). Bergson innovates by combining metacognitivism with noncognitivism. Instead of focusing on the truth-value of negative perceptual reports, Bergson emphasizes their role in expressing and evincing emotions.

According to Bergson, an admonitory speech act is performed with 'The table is not white'. To deny is to do something—to caution against a potential error.

Denials also express feelings. Bergson focuses on regret (1911, 311). A bride laments, 'The table is not white' because she ordered a white table. This suggests that the perceivability of absences varies with the same factors that increase regret. Regret is most intense for unusual choices that could have easily been decided the other way. Regret correlates with suffering, intimations of mortality, leisure for reflection, imminence, and irreversibility of outcome.

Richard Gale (1976, 59) suspects that Bergson has confused negation with denial. Denial of p is a speech act that is logically independent of the truth-value of p. In addition, denial entails the existence of someone (the denier). Logical laws that are plausible for negation (double negation, excluded middle) are implausible for denial.

Since Bergson wrote before J. L. Austin's (1975) distinction between illocutionary acts and perlocutionary effects, the charge sticks. However, we can salvage more of what he says by seeing him as an early expressivist about absences. The subjectivist says that 'X is absent' means 'There is a frustrated expectation of X'. The expressivist denies that 'X is absent' reports anything. Instead, it is a sigh of disappointment concerning X. Whereas a report can be true or false, sighs have no truth-value.

Grammar is an incomplete guide when separating genuine perceptual reports from the pseudoperceptual report of negative things. The train conductor announces, 'Watch for the gap as you exit the train'. That seems analogous to 'Watch for the strap as you exit the train'. But there is no such thing as a gap.

According to Bergson, gaps are as impossible. Reality is everything there is. Reality does not include what is not.

Whereas solitary creatures that live in the present can make positive judgments, negative judgments are for social creatures that remember the past and have expectations about the future. Negative "perceptual" reports rely on memory and inference or social skills. They are not based solely on the sense organs.

A total amnesiac could not make negative judgments (1911, 297). He would lack the premises about the past needed to base an expectation. Absent memory, there are no absences.

Negative "perceptual" reports also make essential use of our language faculty (1911, 308). For the function of negative judgment is communication. A total aphasic would not be capable of negative judgments.

Genuine perception relies solely on the senses. That is why loss of memory, reasoning, language, and social skills are not visual handicaps. Thus our apparent perception of negative things cannot be genuine perceptions.

The curious Parisians saw only a positive reality, such as the pegs on the dusty wall. When they said, "I see the absence of the *Mona Lisa*!" they expressed regret.

A contemporary expressivist might allow that at least some of the Parisians were simply lamenting the *Mona Lisa*'s absence and were not aiming at any future action. "Solon, when he wept for his son's death, and one said to him; *Weeping will not help*; answered, *Alas, therefore I weep, because weeping will not help*" (Bacon 1859, 113). News of the end of the world would be lamentable even though the lamentation would not aim at improving the future.

Although I have categorized Bergson as an expressivist, he resembles the American pragmatists in his tendency to look forward to the practical upshot of what we say. The lamentations about the *Mona Lisa* did pressure the police to investigate vigorously. In addition to the usual suspects, the detectives cast a curious eye on the modern art movement. That iconoclastic movement was supported by Bergson's philosophy of creativity. Classical art, epitomized by the *Mona Lisa*, was static and was being overtaken by photography. Real life is actually dynamic, perspectival, and nonmechanical.

Detectives hypothesized that the theft of the Louvre's crown jewel was a vengeful prank intended to humiliate the curators who marginalized modern art. The animus against the art establishment was intense. Pablo Picasso's promoter, Guillaume Apollinaire, had signed a manifesto threatening to burn down the Louvre. Apollinaire possessed two stolen statues from the Louvre—which Picasso had studied for the *Demoiselles d'Avignon*, a defining painting of modern art. After the two revolutionaries abandoned a plan to toss the statues in the Seine, Apollinaire turned in the statues to a newspaper editor for fear of being arrested for the theft of the *Mona Lisa*. This compromise backfired. Apollinaire was arrested and pressured to implicate Picasso. Picasso only avoided imprisonment by disavowing a significant connection with Apollinaire. Picasso's lie eliminated himself as a suspect and so helped redirect the police toward the true culprit. In 1923 interview for *Art News* Picasso defined art as a lie that tells the truth.

Bergson agrees with the American pragmatists that action takes priority over thought. However, Bergson does not associate truth with usefulness. Utility is instead a source of illusion.

According to Bergson, we think to act. Action is designed to fill the gap between desire and reality. Our attention narrows to this gap, and we neglect the positive background needed to host the gap. This void becomes a blank canvass upon which we must embroider a positive reality. This habit of concentrating on gaps makes nothingness seem like the default state. Positive reality becomes the thing that needs explanation.

And where will the explanation come from? From the realm of necessary truths. Thus the presupposition that the empty world is possible encourages Plato's conception of reality in which the mark of the real is permanence. Since Bergson regards reality as dynamic, he needs to challenge the presupposition that pure nothingness is possible.

Language is a servant of thought. Consequently, there is a chain of command descending from action to thought and from thought to language. Action's preoccupation with gaps is reflected in the language of negation and nonexistence.

On the surface, 'Troy does not exist' appears to be the affirmation of a negative, reality just as 'Troy exists' is an affirmation of a positive reality (1911, 312). But the negative existential 'Troy does not exist' is actually a relational statement. It alludes to a positive reality that precludes the truth of 'Troy exists'. 'Troy does not exist' is actually a double affirmation. The first affirmation is the admonitory metajudgment: 'Any judgment that Troy exists would be mistaken'. The second affirmation is the indeterminate generalization that there is some positive state of affairs that prevents Troy from existing.

Bergson's attack on the perception of nothings is intended to undermine any empirical ground for believing in negative things. But he is by no means agnostic about them at a metaphysical level. Bergson thinks that negative things are impossible.

Talk of relative absences can be paraphrased into metacognitive remarks. But metaphysicians fall into incoherence by misconstruing these relative absences as absolute absences. We cannot form a mental image of absolute nothingness. Since talk of nothingness is not backed by ideas, 'Why is there something rather than nothing?' is a meaningless question.

We can experience local annihilations. But these are just substitutions of one positive thing with another. If the *Mona Lisa* were burned, the portrait would be turned into ashes, heat, and gases. Each of those is something—just something less preferred.

Annihilation is not a disappearing act. Any awareness of nothing would be self-defeating: "At the very instant that my consciousness is extinguished, another consciousness lights up . . . it had arisen the instant before in order to witness the extinction of the first" (Bergson 1998, 278).

Whenever there is talk of annihilation, we should look for hidden relata. Specifically, annihilation always requires an annihilator and a substitute for what had earlier existed.

'Why is there something rather than nothing?' falsely presupposes the possibility of total nothingness. The presupposition is sometimes defended with the subtraction argument: First imagine a world with three objects, then a world with two objects, then one, and then zero objects. Voilà! The empty universe.

Bergson objects that this recipe commits the fallacy of division. From the premise that anything can be subtracted it does not follow that all can be subtracted. A casserole can survive without any particular ingredient. But the casserole cannot survive without any ingredients.

Total nothingness would require an absence of time. But consciousness of absence requires belief in the existence of time: "There is absence only for a being capable of remembering and expecting. He remembered an object, and perhaps expected to encounter it again; he finds another, and he expresses the disappointment of his expectation . . . by saying that . . . he encounters 'nothing'" (Bergson 1911, 281).

Actuality has primacy over possibility. To think something is merely possible is to think something else blocks it from coming into existence. An empty world is impossible because there would be nothing blocking things from coming into existence. Each thing has inertia toward being.

Bergson's fundamental intuition is that reality is positive. Any negative truth must be an indirect positive truth. For instance, seeing holes in a page must be interpreted as seeing a perforated page—where this is understood as a particular shape.

The Skeptical Parallel between Absences and Causes

There is a parallel between Bergson's skepticism about the perception of absences and David Hume's skepticism about the perception of causation. According to Hume, all we witness is a succession of events. Causation is inferred as a force working behind the scenes.

Albert Michotte (1946) studied conditions under which one ball will appear to chase, lead, or force another ball to move. Subsequent psychologists made videos that elicit causal attributions. They touted the films as evidence that causation can be perceived. But Hume would cite the movies as striking evidence *against* there being causation out in world. The viewer only sees frames of a movie, none of which bears a causal connection to a successor. All the psychologists have furnished is further evidence that causal attribution can be triggered without any underlying causation.

Instead of justifying confidence in a future event, the attribution of causation expresses confidences—and attempts to elicit that confidence in others. In addition, it sets the groundwork for praise, blame, and remediation. Since we have different roles, search for "the cause" will vary. The police, the weatherman, and the road engineer nominate different causes for the automobile accident. As evident from discussion of omissions, these factors will also influence negative judgments.

Hume takes the further step of denying that there is any underlying causation. Correlations breed habitual response—in this case the prediction that the same type of event will occur as did occur. The perceiver projects this force of habit on to the external phenomena. Similarly, Bergson could argue that we project our negative feelings of frustration on to the external world. The angry golfer kicks the hole that has spoiled his game.

According to Bergson, consciousness is designed to see the cup as half empty rather than half full. For consciousness serves desire and desire can only be served by fixing what is wrong, not savoring what is right.

So who stole the *Mona Lisa*? An Italian employee of the Louvre, Vincenzo Peruggia, eventually admitted that he removed the small painting when the salon was empty. He just unhooked the painting from the wall, took off his white smock, and walked out through a stairwell. After storing the *Mona Lisa* in his apartment for two years, he returned to Florence. According to Peruggia, the real thief was Napoleon. Peruggia simply wanted a reward from the Uffizi Gallery for the return of stolen property. The director of the Florence gallery agreed. Secretly, he alerted the police. Peruggia was arrested when he tried to collect his money. The court agreed with the Italian public that Peruggia's motive was patriotism. The judge leniently sentenced Peruggia to a year and 15 days in prison. He was released after seven months.

Since Peruggia was convicted of the theft, there is a legal answer to the question "Who stole the *Mona Lisa*?" But the court conceded that Peruggia intended his act to be restoration of stolen property, not theft.

Was Napoleon the real thief? No. The *Mona Lisa* was not among the booty collected by the general during his conquests in Italy. In the sixteenth century, Leonardo da Vinci wished to become court painter. He took the *Mona Lisa* to France as a gift for King Francis I.

"The greatest art theft of the twentieth century" was not a great theft in terms of technique. Nor was it great in terms of the object stolen. Before 1911, the *Mona Lisa* had been graded as a fine portrait by a fascinating genius. But it was a fine portrait among a hundred fine portraits. Only after its absence did the *Mona Lisa* acquire its current presence.

21
Sartre
Absence Perceived

Autobiographers commonly begin before their own beginning. But in *Words*, Sartre dismisses his family tree as irrelevant. Like John Scotus Eriugena's God, Sartre negates himself out nothing: "I have never seen myself as the happy owner of a 'talent': my one concern was to save myself—nothing in my hands, nothing in my pockets—through work and faith" (1964, 158). Like Bradwardine's God, absence of determination makes Sartre free. Like Maimonides's God, Sartre has no essence.

Each of us emerges from an infinite past of nothingness, lives, and dies, to face the infinite oblivion that so horrified the Egyptians. Sartre breaks with Martin Heidegger, who emphasizes the temporal asymmetry of birth and death. For Heidegger, we live forward from birth. I could not have born earlier, but I could have died later. For Heidegger, my failure to have been born earlier is not a deprivation, but my premature death is a deprivation.

Autobiography differs from biography in not including the subject's death. Sartre's death merely removes meaning from his life by undercutting all of his projects. When an author completes his story, he is there to view the story in retrospect. The autobiographer is not present. Death is a limit on his experience, not an event, as it is for a biographer.

Fatherlessness Precedes Godlessness

Honeybees have grandfathers but no fathers. Males are produced by queen bees who refrain from fertilizing the egg. The male is a clone of the queen except for gender.

Sartre's preoccupation with absence began with the death of his father, the naval officer Jean-Baptiste. This occurred when baby Jean-Paul was only 15 months old. He later characterized the absence of his father as liberating: "Jean-Baptiste's death was the great event in my life: it returned my mother to her chains and it gave me my freedom" (1964, 15). For Sartre, fathers necessarily interfere with the freedom of their developing children. Good fathers are impossible: "It is not the men who are at fault but the paternal bond which is rotten. There is nothing better than to produce children, but what a sin to have some! If he had lived, my father would have lain down on me and crushed me" (1964, 14–15).

Philosophically, the absent father became a model for the absent God. Sartre became the sort of atheist who experiences the absence of God as a gaping metaphysical hole. Sartre puts the hole in holy.

Bergson's Influence

Intellectually, Henri Bergson is a father figure for the young Sartre. But instead of filling the void, Bergson precludes the void. Reality is what is, not what is not. Given Bergson's expressivism, absences exclaim but never explain.

Behind every negative is a positive. 'There is not a female pope' is made true by a positive fact such as the Catholic Church's regulation that all priests be men and the practice of drawing popes from the priesthood. Once we have the positive facts and the notion of negation, we can derive all the negative facts. 'There is nothing' would be a contingent, negative fact. But then it would have to be grounded on some positive reality. That positive reality would ensure that there is something rather than nothing.

In an early book (1940) *The Imaginary* Sartre supports Bergson with the principle that every thought requires something to think about. One cannot simply think. One must think *about* something.

> There cannot be an intuition of nothingness, precisely because nothingness is nothing and because all consciousness—intuitive or not—is consciousness of something. Nothingness can be given only as an

> infrastructure of something. The experience of nothingness is not, strictly speaking, an indirect experience, but is an experience that is, on principle, given "with" and "in." Bergson's analyses remain valid here: an attempt to conceive death or the nothingness of existence directly is by nature doomed to fail. (Sartre 2004, 187)

Sartre agrees that one can conceive of one's body as a corpse. But to directly conceive of death, one would have to imagine lacking any consciousness at all. This is impossible. Your own death is inconceivable.

Further study of Edmund Husserl gave Sartre second thoughts about whether the principle of intentionality really precludes thoughts about nothing. In *Being and Nothingness*, Sartre tacitly targets Bergson's restriction of negation to metajudgment:

> Furthermore ordinary experience reduced to itself does not seem to disclose any non-being to us. I think that there are fifteen hundred francs in my wallet, and I find only thirteen hundred; that does not mean, someone will tell us, that experience had discovered for me the non-being of fifteen hundred francs but simply that I have counted thirteen hundred-franc notes. Negation proper (we are told) is unthinkable; it could appear only on the level of an act of judgement by which I should establish a comparison between the result anticipated and the result obtained. (1969, 6)

Sartre goes on to argue that absences can be the objects of judgments (not just metajudgments). A physicist can think about monopoles while being completely agnostic about whether monopoles exist. He can as easily suppose that monopoles do not exist and as suppose that they do exist.

Admittedly, perception is a more partisan mental state than the neutral attitudes of supposition. One can see only what exists. So even if Husserl shows that we can suppose what does not exist, this would not be enough to show that we can *perceive* what does not exist.

Sartre is always loyal to Bergson's principle that positive reality is the substrate for everything. One of Sartre's motives is his activist ethics. Folk morality and law agree that there is a morally relevant difference between acts and omissions. Letting someone die is less grave than

Figure 21.1 Upside-down *Mona Lisa*

killing even though the consequences are the same. Sartre rejects this asymmetry. To not act is to act; omission always involves some substitute action. When a bystander *chooses* not to get involved, he *walks* on, *averts* his gaze, *distracts* himself.

Metaphysically, Sartre differs from Bergson in believing that virtually everything people think about concerns institutional facts. The perception of an artifact requires extensive construction. To see the *Mona Lisa*, one must organize the mosaic of color patches into a coherent whole. The human figure must be separated from the landscape. Negative difference making, contrasting figure from ground, is a prelude to positive construction.

This figure-ground perceptual holism explains why there is an optimal viewing distance for paintings. It also explains the relevance of orientation. Distortions of the *Mona Lisa* pass unnoticed when the picture is presented upside down (fig. 21.1).

In a gestalt switch, the stimulus is constant but the perception varies (fig. 21.2).

In addition to illustrating variation between observers, gestalt switches reveal variation over time with the same observer. There are two intentional objects but just one stimulus.

Gestalt switches substantiate the phenomenological difference between seeing a scene in a purely positive fashion and seeing it

Figure 21.2 House-wine glass array

as having absences. In *Being and Nothingness* Sartre describes his tardy arrival for an appointment with Pierre. Sartre eyes each customer in the cafe: *Êtes-vous Pierre*? Each candidate for being at the forefront of Sartre's attention is defeated. This first wave of negations forms the ground. If Pierre were spotted, he would pop out from the crowd.

> But now Pierre is not here. This does not mean that I discover his absence in some precise spot in the establishment. In fact Pierre is absent from the whole cafe; his absence fixes the cafe in its evanescence; the cafe remains ground; it persists in offering itself as an undifferentiated totality to my only marginal attention; it slips into the background; it pursues its nihilation. Only it makes itself ground for a determined figure; it carries the figure everywhere in front of it, presents the figure everywhere to me. This figure which slips constantly between my look and the solid, real objects of the cafe is precisely a perpetual disappearance; it is Pierre raising himself as a nothingness on the ground of the nihilation of the cafe. So that what is offered to intuition is a flickering of nothingness; it is nothingness of the ground, the nihilation of which summons and demands the appearance of the figure—the nothingness which slips as a *nothing* to the surface for the ground. It serves as foundation for the judgment—"Pierre is not here." (1969, 42)

Sartre sees the absence of Pierre in a cafe. Nothingness pops out of Being. But this hollow pit is always shaped and situated by expectation. If Sartre's companion Simone does not anticipate Pierre, she will be presented with exactly the same cafe scęne. Yet she will see it differently.

Why does Sartre not also see the Duke of Wellington? Because Sartre was not anticipating a rendezvous with the duke. This psychological explanation gives hope of answering Aristotle's proliferation challenge: if there is one absence, why not millions?

If Simone had expected the Duke of Wellington, then she would have seen the absence of the duke. The relativist about absences would say that the duke is *absent for Simone* but not *absent for Sartre*. For the relativist, all absences are *absences for someone*.

Sartre is no more a relativist about absences than he is a relativist about pains. Headaches depend on sufferers but are not relative to sufferers. When Sartre visits an unfamiliar classroom, he knows that the teacher knows who is absent even though Sartre does not know who is absent. The students are not merely *absent for the teacher*.

Absences track the most informed perspective. Consider the joke: Jean-Paul Sartre sits in a cafe, writing *Being and Nothingness*, when a waitress approached him: "May I get you something to drink, Monsieur Sartre?" "*Oui*," replies the French existentialist, "*Une tasse de café* with sugar, but no cream, *s'il vous plaît*." A few minutes later, the waitress apologizes, "I am sorry, Monsieur Sartre, we are all out of cream—how about with no milk?"

Sartre writes in the manner of a Cartesian dualist, strongly contrasting the inertness of objects ("being in itself") with consciousness ("being for itself"). Mescaline breaks down the contrast. Objects awake. When the experience is positive, tables and brooms are friendly, inviting an I-thou relationship rather than an I-It. When negative, they menace, like a dog that turns on you, and becomes Monsieur Dog.

Sartre would joke about how bad experience with mescaline led him to recoil from the vegetation that creeps toward the city. The humor lays in an anti-panpsychic contrast with reality. Contrary to Schopenhauer, objects are only passively influenced by consciousness, in the way pieces of papers become train tickets through human fiat. The conductor must be wary that he not be defined in the same way.

Sartre says all potentiality and all destruction depend on consciousness. But this is not an expression of immaterialism. Sartre's intent, as stated in his *Critique of Dialectical Reason*, is to be a materialistic monist—which is reinforced by his sympathy for Karl Marx. Sartre regards himself as no more a dualist than the ancient atomists. Leucippus says that being (atoms) and nonbeing (void) are equally real. But just as nothingness plays nonmaterial roles in atomism (explaining unnatural motions such as water rising above air through a siphon), nothingness plays nonmaterial roles in Sartre's metaphysics (blocking Freudian causal chains that threaten freedom, distinguishing people from mere objects).

The difference between Sartre and the atomists can be brought out with James McTaggart's distinction between two temporal systems. The A series (yesterday, today, tomorrow) is organized from a perspective and is crucial for action. The B series is objective; it just assigns coordinates in space and time: Paris (latitude 48°52′0″ N, longitude 2°20′0″ E), August 18, 1903. The atomists were interested in nothingness as analyzed in the style of the B series. Sartre is interested in nothingness as analyzed in the perspectival style of the A series.

The Observer Relativity of Absences

According to Pierre Duhem, theories entail predictions only when conjoined with auxiliary assumptions. This explains why the blame for a failed prediction can be deflected from a theory to background assumptions. If absences are refuted predictions, then we should expect them to depend on both theory and background assumptions. People with different theories and background beliefs should vary in what absences they perceive.

Art historians focus on the *Mona Lisa*'s complete absence of jewelry. She has no rings, no bracelets, no necklace. The historians know that the norm for the era was to paint subjects with expensive adornments.

Although the *Mona Lisa* also lacks any representations of eyelashes, some art historians deny that Leonardo depicted her without eyelashes. According to them, the depiction of absences requires expectation to be raised about presences. For instance, painting only the right eye with lashes would have constituted a depiction of her left eye as devoid of lashes. Expectations of detail vary with genre. Stick men figure are so sketchy that there is little scope for the depiction of absence.

Eye*brows* are obligatory for formal portraits. Since there are no visible strokes for *Mona Lisa* eyebrows, some historians concluded she is represented as without eyebrows. But recent evidence supports the conservative theory that the strokes have merely become invisible with the passage of time (high-resolution photographs show traces of slight eyebrow lines).

Expectations can also be manipulated against our will. Marcel Duchamp (1919) first vandalized *Mona Lisa* in the obvious way by drawing a mustache (fig. 21.3).

He then engaged in invisible graffiti by presenting an untouched postcard with the title *Mona Lisa Shaved*. Duchamp's motto: "The *spectator* makes the *picture*" (Judovitz 1987, 187) Even against one's will! It is difficult to unsee *Mona Lisa*'s mustache.

Expectations are built out of propositions. And propositions are built out of concepts. If pigeons lack the concept of a mustache, then Duchamp cannot make the pigeon see *Mona Lisa* as having a mustache. Therefore, Duchamp cannot make the pigeon see *Mona Lisa*'s absence of a mustache.

Admittedly, pigeons would have trouble recognizing the absence of a mustache even if they had the concept of a mustache. Pigeons are very slow to learn to solve any problem involving recognition of an absence. Four-year-olds are almost as bad (Sainsbury 1973). Even adults exhibit the "feature-positive effect": far more trials are needed to recognize an *absence* of a feature as being a key difference.

What goes for discrimination learning goes for memory and perception. This positive-negative asymmetry is predicted by Henri Bergson's thesis that negative judgments are metacognitive. Human beings have far stronger metacognitive abilities than other animals. Metacognition matures gradually, exhibits considerable interpersonal variation, and is cognitively taxing.

Figure 21.3 Duchamp's vandalized *Mona Lisa*

Conceptually Dependent Absences

The spiritual nature of air led the mathematical philosopher Pythagoras to ban beans (to prevent *a posteriori* transmigration). Cicero (1923, 58) concludes, "Somehow or other no statement is too absurd for some philosophers to make."

But there is an historical limit to credibility. Those preceding Leonardo could not assert that Leonardo's *Mona Lisa* depicts a woman with a mustache. The pre-Leonardos could not believe the absurdity because they lacked access to the vocabulary needed to express the proposition. To understand a name, one must have a causal connection with its bearer. This was not possible for Leonardo's predecessors because 'Mona Lisa' names an object that had yet to exist.

The historical limit extends to predicates such as 'helicopter'. The ancient Egyptians have drawings that resemble helicopters. But they cannot be depictions of helicopter. Ancient Egyptians lacked any appropriate connection to these artifacts.

These connections need not be causal. Inventors of the helicopter had the concept of a helicopter before there were any helicopters. All they needed was a sufficiently detailed description of a helicopter.

There is vagueness as to what constitutes sufficient detail. Leonardo da Vinci's design for a helical air screw is sometimes touted as the invention of the helicopter. His design from the Codice Atlantico embodies insights into principles relevant to helicopter flight: "I have discovered that a screw-shaped device such as this, if it is well made from starched linen, will rise in the air if turned quickly" (Leonardo 1978, folio 83v of manuscript B). But many more insights and technological advances would be needed for a helicopter that could really fly.

"Seeing as" only requires conceptual competence, not belief. When you learn that some Romans believed that the stars were holes in the celestial sphere, you can see the stars as holes just as the Romans. But Romans cannot see the stars as you do—as giant bodies undergoing nuclear reactions. You are their conceptual superior. In turn, your grandchildren will be your conceptual superiors. They will spot absences in your landscape that are necessarily invisible to you.

These reflections on "seeing as" can be a useful retreat for Sartre. He claims that the perception of absences requires frustrated belief. But this is over-opinionated. A detached astronomer can see a shadow while suspending judgment about whether he is seeing a shadow or a hole or a black rock. Instead of being relative to a belief, Sartre should say that absences are relative to a question. He comes close to this better basis for absences at the beginning of *Being and Nothingness*. But the interrogative mood is a poor fit with Sartre's assertive personality.

Fictionalism about Absences

The fictionalist says that Sartre is merely pretending to see the absence of Pierre. When someone pretends to see a ghost, he goes through the motions of witnessing. But there is no genuine observation.

Sartre rejects fictionalism on the grounds that it fails to provide relief from the paradoxes of negative cognition: "To call this conduct pure fiction is to disguise the negation without eliminating it. 'To be a pure fiction' here is equivalent to 'to be only a fiction'" (1969, 5).

Bergson analyzes negative statements as the positive assertion that there exists something that *precludes* the truth of the negated statement. 'Preclude' is a disguised negation.

Sartre's logical objection to fictionalism can be supplemented with developmental psychology. Even young children distinguish between seeing an absence and pretending to see an absence. The psychologist Alan Leslie (1994, 223) hosted the most famous tea party in experimental psychology. Children were encouraged to "fill" two toy cups with make-believe tea. The experimenter then says, "Watch this!" and pretends to pour out the tea from one of the cups by turning it over and shaking vigorously. The child was asked which cup was full. Ten of twelve pointed to the cup that the experimenter had not touched.

The children knew that both cups were really empty. Absences, therefore, cannot be a matter of pretense. Pretend emptiness corresponds to pretend bankruptcy in a Monopoly game. The money is pretend money and the bankruptcy of the loser is pretend absence of money. All of this is compatible with money being a creature of expectation. What makes something money is its power of exchange—power that must analyzed in terms of the expectations of the money exchangers.

Sartre resembles a fictionalist because he closely associates absences with negation. But pretense only roughly correlates with negations in the logical sense. True, one typically pretends that p while believing that not p (as when the child pretends that a cup is full, while believing it not to be full). But the child can also pretend that p while believing that p (as when the child both pretends that the overturned cup is empty and believes it to be empty).

The Multilocation of Absences

After the *Mona Lisa* was stolen, Parisians were reported as viewing the empty space where the *Mona Lisa* had been. Was the absent *Mona Lisa* identical to this empty region? The standard objection to identifying absences with regions of space is that absences can move. Regions of space are necessarily where they *were*, where they *are*, and where they *will* forever be.

Sartre's psychological approach suggests a second objection to identifying absences with regions of space. Unlike an empty region, the *Mona Lisa* can be absent in two places at the same time. For suppose the painting is overbooked at two exhibitions and fails to appear at either museum. Unlike a physical object and unlike an empty region of space, an absence can be multilocated.

What can be bilocated can be trilocated (just add a third, expectant curator). In principle, expectations can be multiplied so that the absence of *Mona Lisa* became omnipresent. Less hypothetically, young Sartre assumes God to be omnipresent. Upon conversion to atheism, Sartre experiences an omnipresent absence.

As creatures of expectation, absences depend on us for their existence. This parasiticism does not rob them of significance. The fact that pain depends on a sufferer does not diminish its significance. Showing that money depends on expectations did not console Isaac Newton when the South Sea Bubble burst. The psychological analysis of absences does not demote their importance. Indeed, once one realizes that most of what we care about is psychological, Sartre's analysis tends to fortify them from attacks as mere nothings.

Sartre's Compromise

Sartre agrees with Bergson that being has priority over nothingness. Absences are always particular absences whereas presences can be general. "In a word, we must recall here against Hegel that being *is* and that nothingness *is not*" (1969, 15).

Against Bergson, Sartre says that negative perceptual reports are much as they appear. According to Sartre, 'The table is not white' is a genuine, perceptual report.

Yet Sartre agrees with Bergson that negative perceptual reports are based on frustrated expectations. Positive reality is what gives rise to perceivers and so is more fundamental than negative reality. The slogan "To be is to be perceived" is false. But its negative variation, 'To be absent is to be perceived as absent', is true.

Lastly, Sartre accepts Bergson's anthropocentrism about absences: "Non-being always appears within the limits of a human expectation" (1969, 7). This anthropocentricism is natural for Bergson because he interprets denials as complex, higher-order, social acts that require language. Sartre has lowered the bar for the perception of absences. Consequently, his humanism is more embarrassed by our continuity with animals.

Epistemic versus Nonepistemic Seeing

Sartre's compromise has a translation into analytic philosophy that suggests a principled way of segregating human perception from animal perception. The first draft of this translation is written with the distinction between epistemic and nonepistemic seeing. Epistemic seeing, "seeing that," depends on belief. This is seeing as a sensory type of belief pickup. There is always some full proposition that goes with epistemic seeing.

Nonepistemic seeing does not involve belief in any distinctive way. A dog sees a fire hydrant without believing that it is a fire hydrant. Although we often puzzle about what beliefs to ascribe to animals, we confidently attribute vision to them. Even bees see. But we are not confident bees believe. Bees resemble neurons. Neurons register information that may contribute to a belief. But the neuron is itself unopinionated. There must be a cognitively less demanding sense of 'see' that does not entail belief.

Nonepistemic seeing is a matter of looking in a sufficiently informative direction with a functioning visual system and with adequate uptake by the sensory system (uptake that can fall short of consciousness). Nonepistemic seeing is automatic. It is a necessary prelude for *seeing that* something is the case. The ancients saw objects undergoing nuclear reactions in the night sky. But they did not see *that* these objects were undergoing nuclear reactions.

If observation were completely theory dependent, experiments would be circular and private. What you see would be subservient to what you believe you see. Although the history of science contains many episodes in which theory has dominated what counts as an observation, there are also reassuring examples of observation overcoming theory.

In the terminology of analytic philosophy, Sartre's compromise is that all seeing of absences is epistemic seeing. Whereas Bergson denies that there is any observation of absences, Sartre affirms that there is limited seeing of absences. All perceptions of absences are at the level of epistemic seeing. This by no means demeans them. Sartre thinks that most of what is perceptually interesting depends on consciousness.

Richard Taylor's Objectivism

Richard Taylor (1952) tried to demonstrate that there is nonepistemic seeing of absences with the diagram in figure 21.4.

You see the absence of a dot in the second circle. Taylor concludes that negative things exist.

But the right circle equally has an absence of dashes and tildes. According to Sartre, we *infer* these absences from what we do see (which is entirely positive). We only see absences indirectly, as we see wind. Just as we infer wind from a billowing sail, we infer an absence of a dot from seeing a positive feature of encircled area. Absences are theoretical entities that require the sort of intellectual sophistication Louis

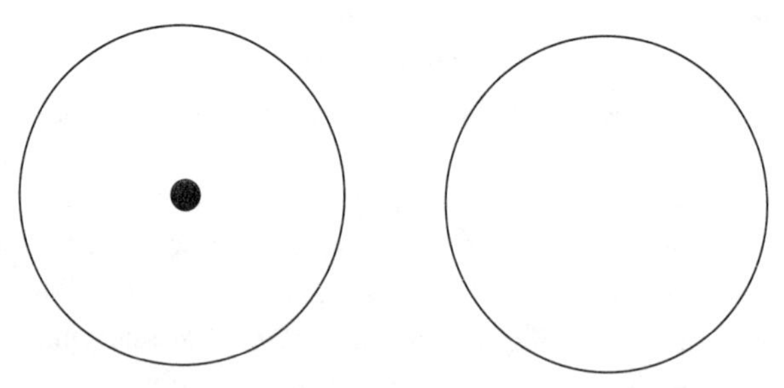

Figure 21.4 Taylor's Two Circles

Pasteur exhibited when inferring the existence of germs. The only way we can justifiably believe that there are absences is by their explanatory power.

According to the analytic Sartreans, the curious Parisians wished to see that the *Mona Lisa* was absent *from the Louvre*. They had to travel to the Louvre for this experience.

The objectivist counters that the absence of the *Mona Lisa* did not depend on expectation. The *Mona Lisa* was already absent before anyone missed it. When the curious Parisians queued to see the absent *Mona Lisa*, they already knew the *Mona Lisa* was absent from the Louvre. They were seeking an antiquarian thrill of directly perceiving the absence—the same antiquarian thrill one gets by seeing the *Mona Lisa* when it is present.

There was a rumor that the *Mona Lisa* had long ago disappeared from the Louvre and that a *copy* had actually been stolen. If that rumor were true, then the crowd would have seen the absence of the copy of the *Mona Lisa*—not the absence of the *Mona Lisa*. For the absence depends on the host object, not beliefs about the host object.

The identity of absences is controlled by causality rather than resemblance. When a single twin is missing from a class, the teacher sees the absence of only one of the twins. Just as she does not need to be able to distinguish between the twins to see one twin rather than the other, the teacher does not need to be able to distinguish between their absences to see the missing twin's absence.

For the objectivist, seeing the *Mona Lisa* is like holding Leonardo's paintbrush. Direct contact gives us knowledge *by acquaintance*. No amount of propositional knowledge (feeling *that* the object in one's hand is Leonardo's paintbrush) gives us acquaintance knowledge.

How does one have direct contact with an absence? The objectivist answers by appealing to ordinary standards of contact. Doubting Thomas feels the hole made in Jesus's body by sticking his finger in the wound. Running a finger over the lining of a hole constitutes feeling the hole (even if misfelt as a bump).

Unlike shadows and silence, holes can be perceived by more than one sense. Holes can be both felt and seen. Arguably, holes are also audible. Punctures in bicycle inner tubes are easier to hear than see.

If the invisibly punctured tube is inflated and then immersed in water, bubbles will allow the repairman to indirectly see the hole. When he focuses on the spot, the repairman can then see the hole itself. There is a natural contrast between indirectly seeing absences and directly seeing them.

Startle Reflex

The objectivist's strategy is to dumb down absences. As the diving bell descends, the pressure becomes too great for the subjectivist to follow. Consider our most primal reaction to absence. The Moro reflex in infants is stimulated by the sudden loss of support. The baby's arms stretch out, then contract, and finally the baby cries loudly. Like other reflexes, it requires no learning and is not extinguished by habituation to the stimulus.

Prey fish have a startle reflex toward looming shadows (as when black cardboard is passed over the surface of an aquarium). Nile tilapia undergo cardiac arrest. These fish are exposed to aerial predators. The sudden absence of light is a sign of mortal danger. The fish reacts too quickly for inference to play a role.

Startle is an invariant, all or none response that happens too quickly to wait for consciousness (Robinson 1995). (The speed makes startle responses too fast to accurately fake.) So the absences associated with startle responses cannot be judgment dependent.

Externalist Anomalies

The objective aspects of perception pose riddles that cannot be answered by retreat from belief to seeing-as.

First, perceptions of absences are corrigible. Pierre could be behind a pillar. No amount of psychic fizz will conjure the absence of Pierre into existence given that Pierre is present in the cafe. There can be no détente between Pierre and his absence.

Second, absences are not self-intimating. If Sartre sees Pierre's identical twin at the cafe then Sartre will *feel* no disappointment.

Nevertheless, Sartre's expectation will not be fulfilled. Pierre will be absent but will not be experienced as absent. Amputees commonly fail to detect their absent limbs. A phantom limb hallucination masks the absence.

One might complicate Sartre's theory by requiring that there be a match between the representation of an absence and reality. The worldly side of this match will be problematic. For the whole appeal of the psychological theory is that it avoids commitment to negative states of affairs.

Looking at Absences without Noticing Them

An absence of X can arise without anyone even raising a question about X. Consider the phenomena of unnoticed absences. In the summer of 1911, Louis Béroud had been copying the *Mona Lisa* in the Salon Carré. On August 22, he arrived in the morning to complete his copy. Béroud instantly spotted the absence of the *Mona Lisa*. He asked the guard where it was. The guard had not noticed the *Mona Lisa*'s absence.

Regaining his composure, the guard surmised that the *Mona Lisa* was off being photographed. But the guard did not find it in the photography department. Nor was it found anywhere else in the Louvre. The *Mona Lisa* had actually been stolen the day before.

The police were suspicious of the guard. The guard must have looked at the wall where the *Mona Lisa* was hung. How could the guard have missed it?

Experiments on change blindness suggest that the guard was simply inattentive. Even if he looked directly at the spot, he might fail to see that the painting was absent.

In cases of inattentional blindness, people look at the object in an appropriate way but fail to register it. Motorcyclists who have been struck by automobiles often recall that the driver looked right at them but just kept on driving. The problem is attention rather than visibility. If the driver is looking for other automobiles, then he is less apt to notice a motorcyclist. This attention error is made more likely when there are competing demands for the driver's attention.

Change blindness differs from inattentional blindness in being a failure to update. After laboratory subjects see a photograph of a plane with soldiers boarding, they presume the engine will be there when they next look at the photograph. Few of them noticed when a doctored photograph was substituted—a photograph that subtracted the plane's engines.

Another study focused on the sort of continuity errors familiar from the cinema. The researchers made their own short film in which the actress's scarf disappears. Few viewers noticed the disappearance.

The lesson psychologists draw is that vision demands attention. There is only a limited amount to go around. People are vulnerable to "blindness" by distraction. If their attention is taxed, their visual performance will suffer.

Change blindness may be a symptom of the frame problem. When a situation changes, one cannot afford to review every possible ramification. One must operate with a default principle that lets "sleeping dogs lie."

Even attentive search does not guarantee recognizing what one is looking at. A novice fossil hunter can look right at a fossil tooth and not distinguish it from the clutter of pebbles and sand.

Adults are better at visual search than children—despite weakening eyes. Grandparents notice this when reading "Find it" puzzle books with their grandchildren. These puzzles require the child to spot an object in a cluttered field of objects. The grandparents break through camouflage faster than their grandchildren.

Disunified Perceptions

Another objectivist argument for the perception of absences is based on disunified perception. A person can see an absence with one eye and misperceive it as present with the other eye. If the perception of absence occurred at the level of consciousness, then the unity of consciousness would prevent bifurcated perceptions of absences.

The neuroscientist Margaret Livingstone appeals to another disunity to explain the mystery of the *Mona Lisa*'s smile. When you look at Mona Lisa's eyes, she seems to smile. When you check by looking at her lips, she seems to not smile. Livingston's explanation relies on the distinction between foveal and peripheral vision. Foveal vision is proficient at discerning detail. It is less suited to detecting shadows. Peripheral vision has complementary strengths and weaknesses. It is adept at picking up shadows. When we look at Mona Lisa's eyes, the shadows are detected by peripheral vision and are used to construct a smile. When we look directly at her lips, the shadows cannot be detected by foveal vision and do not provide the cues necessary for the perception of a smile.

Imperceptible Absences

If seeing absences were a matter of seeing *that* something is the case, one should be able to see the vanishing point behind Mona Lisa's head. For a trained artist can see that her head occludes the vanishing point. But he cannot see the vanishing point. Removing her head would not help.

The perception of absences is conditioned by its positive counterpart. More specifically, the perception of absences is shaped by counterfactuals (which are in turn shaped history, laws of nature, and resemblances). The particulars of the absence are what would have been perceived. To preserve the "wall of shame," the Louvre's curators did not move the paintings by Titian and Correggio that served as bookends for the *Mona Lisa*. When curiosity dwindled, they ended the spectacle by putting Raphael's *Castiglione* in the *Mona Lisa*'s place and shuffling the order of the surrounding pictures.

This counterfactual principle explains why we do not perceive the absence of bacteria in an autoclave. It also explains one cannot see mathematical absences and psychological absences such as lexical gaps. The range of what is negatively perceivable is determined by what is positively perceivable.

Summary

Sartre began as a disciple of Bergson's metacognitive view of absences. What appear to be perceptions of absences are sophisticated, indirect reports of what is positively perceptible. But a combination of Edmund Husserl's phenomenology and gestalt psychology led Sartre to regard even positive perceptions as highly constructed. Negation is such a central feature in this construction that Sartre concluded that negative perceptions differ only modestly from positive perceptions. Although negative perceptions are filtered through human expectation, they are genuine perceptions. The substrate of reality is entirely positive. But human consciousness builds up a much more elaborate world of constructed entities. This sets up the possibility of perception of absences.

The third position, objectivism, is a reaction to Sartre's compromise. Once his bargain has been translated into the analytic idiom of epistemic versus nonepistemic seeing, Sartre comes off as implausibly cerebral and anthropocentric. The objectivist uses empirical psychology to deintellectualize the perception of absences. To resolve the inconstancy, he permits causation by absences. To stave off metaphysical objections, he refuses to lodge absences on either side of the being versus nonbeing divide. He treats absences as awkward characters, like space and time. It is better to say they exist than that they do not. But the upshot of affirming their reality seems particularly vaporous. The objectivist looks to his feet in embarrassment and then cast his eyes into the distance, hoping for rescue by future generations.

22

Bertrand Russell

Absence of Referents

The Islamic profession of monotheism, 'There is no god but God' highlights the distinction between the predicate 'god' and the name 'God'. To be a predicate atheist is to believe there is no x for which 'x is a god' expresses a true proposition. To be a name atheist is to believe that the purported name 'God' has no bearer.

Ancient Roman critics of Christianity were name atheists but not predicate atheists; they believed that God did not exist but believed that there were gods. Genuine gods were local deities whose worship contributed to civic life. Ancient gods were akin to the royalty of a region—which Roman conquerors preferred to co-opt rather than extinguish.

There can also be predicate atheism without name atheism. Imagine an archaeologist, Kripshe, who infers 'God' was the name of a chieftain who was later regarded as a god. She thinks God is not a god.

Since one individual may bear more than one name, name atheism must be relativized to a set of names. According to Christians, an atheist who denies that God exists but affirms that Jesus exists is a name atheist with respect to 'God' but a name theist with respect to 'Jesus'.

Hindu polytheists accept hundreds of gods. The task of memorizing their names would move from difficult to impossible if a Pythagorean deified each natural number. His pantheon of aleph-0 gods would be dwarfed by a more prolific Pythagorean who regards each real number as a god. When the number of gods is infinite, there must be infinitely many nameless gods. As finite beings, we can only learn a finite number of primitive terms.

Names as Labels

John Stuart Mill held that names are tags rather than descriptions. They have no meaning beyond the item to which they are attached. 'Sue' is reserved for females, but there is no contradiction in a boy being named Sue. Since names lack any meaning apart from their bearers, we do not analyze terms that are classified as names. Names are primitive terms.

Conversationalists presume that names have bearers. This presumption plus primitivism sums to semantic hallucinations. My favorite is from 1936. Democrats had taken control of the Rhode Island legislature and began to award $100 bonuses to veterans. Most Republicans loudly bewailed their rivals' largesse. But one Republican meekly proposed a $100 bonus for Sgt. Evael O. W. Tnesba of the Twelfth Machine Gun Battalion. A delighted Democrat seconded the motion. When the legislature passed the measure, the Republicans erupted in laughter. Democrats were chagrinned to learn that the reverse of Evael O. W. Tnesba is Absent W. O. Leave.

The absence of any descriptive meaning to a name allows reference to succeed without any need to accurately characterize the object. This makes reference as direct as perception—an almost magical form of word power.

If 'God' is a name, then 'God' is an appropriate effect of God. Specifically, for 'God' to be a name, God must be a remote cause of 'God' as used in this very sentence. How might this causal chain have been initiated? Recall how the animals present themselves to Adam for naming (Genesis 2:20). God may have joined the queue. Adam could have dubbed his Creator 'God' thereby introducing a name that could later be used by Eve. She transmits 'God' to her sons. They pass the name to their children. Eventually, your mother whispers 'God' into your ear.

The whispered name would pick out God regardless whether you believe 'God' referred to anything. When Germans correctly pronounce 'Kant' (which rhymes with 'bunt'), titillated American students think this is a gag name. The Americans attempt to join the lewd nonsense, saying 'My professor admires Kant'. What they take to be nonsense is literally true. A crook who thinks he is passing a counterfeit coin might actually be spending a genuine coin—and so be engaged in an unexpectedly legal transaction.

According to Mill, "All names are names of something, real or imaginary" ([1882] 1979, 33). This implies that 'The name "God" has no bearer' is a contradiction. The apparent debate over whether God exists is actually a debate about the nature of God. Some think God is a mind-independent being. Others think God is an idea. Everybody agrees that God exists.

This mentalism converts atheism into a religion. But that is like classifying abstinence as a kind of sex. The atheist denies that 'God' has any bearer. So if a name must have a bearer, as is the case in contemporary classical logic (first-order predicate logic with identity), the atheist must deny that 'God' is a name. It is instead a pseudoname such as 'Saint Ababís' in the Philippines. After the US Army conquered the 333-year-old Spanish colony, Filipinos would hear American soldiers exclaim "San Ababís!" in moments of stress. (*San* denotes a male saint in Spanish as in 'San Francisco' and 'San Diego'.) The Catholic natives believed the soldiers were invoking the patron saint of the United States.

In Anatole France's novel *Penguin Island*, a near-sighted missionary mistakes auks as people—and baptizes them. The Lord rectifies the error by converting the birds into people. Perhaps another intercession could make Saint Ababís the patron saint of nonexistent objects.

When the debunker says 'San Ababís does not exist', he intends his aborted reference to mark the existence of some obstacle in the causal chain (Donnellan 1974). Instead of continuing to a bearer, the chain terminates inappropriately in a mishearing or misunderstanding or misdeed (as in the pseudonames fabricated by liars). 'God' suffers from this type of illusory reference.

Mill was the godfather of Bertrand Russell. Mill's participation in the minting of 'Bertrand Russell' guaranteed that 'Bertrand Russell' was not a counterfeit name. Baptism ceremonies are solemn public events designed to reliably connect a name with the intended bearer and to encourage witnesses to pass on the name from speaker to speaker and from generation to generation. Thanks to a baptism in 1872, Bertrand Russell affects your speech perception by virtue of the use of his name in this sentence. Russell's name is starlight—perceptible even if the star was extinguished long ago.

How much brighter would be the name of God! Hearing his name would be a divine echo. Saying 'God' would constitute direct reference to the Almighty, unmediated by description.

This direct reference would be jeopardized by the sort of garbled usage that led 'Madagascar', though an error by Marco Polo, to shift its reference from a portion of the African continent to the largest island off the African coast. Prohibition against blasphemy can be partly understood as protection of a divine channel of communication.

Direct communication is discouraged by middlemen who advertise expertise in dealing with God. These brokers often advise a total prohibition against uttering the name of God. Some students comply by substituting 'G_d' for 'God' in their essays.

More farsighted clerics found religions with the precaution that their god be nameless. In Palmya there are many altars to an anonymous god. Any name is too familiar a reference to God. Knowing someone's name gives you power over them. It is impious to seek power over a god.

Names of prophets who directly perceived God often inherit divine prohibitions. Section 295C of Pakistan's Penal Code warns, "Whoever by the words, either spoken or written, or by visible representation, or by any imputation, innuendo, or insinuation, directly or indirectly, defiles the sacred name of the Holy Prophet Muhammad shall be punished with the death sentence or imprisonment for life and shall be liable to fine."

Unlike predicates, names spread between languages without need of translation. These highly portable tags do not have any meaning to translate. So names are not included in dictionaries. When you learn the name of the monotheistic pharaoh Akhenaten, you do not learn any Egyptian.

Since a name directly denotes its bearer, a name singles out the same individual at all times. 'Bertrand Russell' still designates Bertrand Russell even though he died in 1970. 'Bertrand Russell' also rigidly designates the same individual in all possible worlds. Consider how we evaluate the subjunctive conditional 'If Bertrand Russell had sat in the nonsmoking section on his October 2, 1948, flying boat plane trip, then he would have perished that day in a Norwegian fiord'. We consider the possible world that most resembles the actual world except Bertrand Russell sits in the nonsmoking section. If he dies in that world on that day, then the counterfactual is true.

In contrast to the rigid designation of names, a description such as 'the youngest co-author of *Principia Mathematica*' picks out different individuals in different possible worlds. In some possible worlds Russell's contemporary Frank Ramsey is the youngest coauthor.

Causal history disambiguates names. There have been several gentlemen with the name 'Lord Russell'. This includes Bertrand Russell after the death of his older brother Frank. In 1959 Bertrand Russell and Lord Russell of Liverpool wrote a joint letter to *The Times*: "Sir: In order to discourage confusions which have been constantly occurring, we beg herewith to state that neither of us is the other."

Are We Born Atheists?

No name can be innately known. Names can only be known through perceptual contact or testimony. We begin with neither. Consequently, innate theism concerns the predicate 'god'. Missionaries were dispatched to save those who had not heard the name of God. Theologians acknowledged that the isolated heathens already possessed the predicate 'god'. After all, the savages were worshiping false gods. Passing on the name of the true god to the wayward would correct their spiritual aim.

Pre-Christian philosophers, such as Plato, showed that study of the predicate 'god' could lead to an improved conception of god. Some theologians thought that this predicate-driven theology could suffice to identify the one and only god. A thinker who never heard the name 'God' could work out that the predicate 'god' applies only to a perfect being. Perfect beings must have all the best properties. Since existence is a better property than nonexistence, all perfect beings exist. But since the property of being all-powerful can be possessed by at most one being, there is exactly one god.

Rationalists admit that a priori beliefs require an infrastructure that only matures after we have had much experience. Innate beliefs in general may require experiences to trigger their manifestation (just as experience of the sun is needed to trigger the unlearned response of tanning). This cause of the belief need not justify belief. For instance, a nearly fatal illness often *causes* the survivor to believe that there is a god. Yet the recovery is not a *reason* to believe that there is a god.

Empiricism began with a denial of innate ideas. According to the founder of British empiricism, John Locke, we enter the world as blank slates. If children are not told of ghosts and other creatures of darkness, they will not fear the dark. The concept of a god must be acquired through experience. For instance, a girl might begin with the idea of her father. She modifies the idea by subtracting limits and grafting on admirable features from other ideas.

Nevertheless, John Locke did not think we are born atheists. Atheism requires belief in the absence of gods, not merely absence of belief that there are gods. A blank slate is an absence of representation, not a representation of absence.

Without the concept of a god, atheism is impossible. The empiricist's best hope for preventing his children from becoming atheists is to prevent them from hearing about gods.

Russell's Conversions

Bertrand Russell's parents were radical members of the British aristocracy. His father, John Russell, Viscount Amberley, wrote an erudite critique of Christianity, *An Analysis of Religion*. He and his wife Katherine were also active in women's suffrage. When John Russell learned that tuberculosis made the family tutor unfit for marriage, Viscountess Amberley was permitted to have sex with him.

As part of a plan to raise his three children as agnostics, Lord Amberley asked his friend John Stuart Mill to serve as the godfather of his second son, Bertrand. Mill was the most prominent of philosopher of his era, and is frequently ranked as the fourth great empiricist (joining the pantheon of John Locke, George Berkeley, and David Hume). Although Mill died the following year, his philosophical views influenced his godson. As a teenager, Russell had been persuaded by the cosmological argument for a necessary being. "At the age of 18, however, I read Mill's 'Autobiography,' where I found a sentence to the effect that his father taught him that the question 'Who made me?' cannot be answered since it immediately suggests the further question 'Who made God?' This led me to abandon the 'First Cause' argument and to become an atheist" (1968, 36).

But a few years later, Russell converted from Mill's empiricism to idealism: "For two or three years . . . I was a Hegelian. I remember the exact moment during my fourth year [as an undergraduate in 1894] when I became one. I had gone out to buy a tin of tobacco, and was going back with it along Trinity Lane, when I suddenly threw it up in the air and exclaimed: 'Great God in Boots!—the ontological argument is sound!' " (1959, 60).

Alfred North Whitehead, hoping to discourage his student's idealism, urged Russell to read Hegel's mathematical writings. This switch from mathematically literate secondary sources to the incompetent primary source exposed Hegel as a charlatan. Or so concluded an embarrassed Russell. The apostate experienced his transfer of allegiance to G. E. Moore's common-sense philosophy

> as a great liberation, as if I had escaped from a hot-house on to a wind-swept headland. I hated the stuffiness involved in supposing that space and time were only in my mind. I liked the starry heavens even better than the moral law, and could not bear Kant's view that the one I liked best was only a subjective figment. In the first exuberance of liberation, I became a naive realist, and rejoiced in the thought that grass is really green, in spite of the adverse opinions of all philosophers from Locke onwards. (Russell 1959, 61–62)

Russell records further conversion experiences. Politically, Russell switched from conservatism to liberalism after witnessing Whitehead's wife in agony. As for marriage, Russell realized—while on a bicycle ride—that he no longer loved his wife. The grammar of Russell's life is punctuated with emotional solecisms.

Philosophically, Russell followed his conversion to naive realism with immersion in Platonic rationalism. He then settled back into an empiricism that was sympathetic to John Stuart Mill's wish to demystify mathematics. Whereas Mill pursued this goal by characterizing mathematical truths as highly confirmed empirical generalizations, Russell reduced mathematical truths to tautologies: 97 + 98 = 195 just means that if you start at position 97 of the number line and advance 98 units further, you arrive at the 195th position on the number line.

Russell's mature view is that name atheism is self-contradictory. Only a name with a bearer is a "logically proper name"—in scare quotes because Bertrand Russell never uses this apt phrase (Proops 2001, 152 fn 2). Words ordinarily regarded as names are actually predicates that apply to exactly one thing if they apply to anything at all. Anselm defines 'God' as 'the best conceivable being'. When Russell denies that God exists, his denial means that either there is no best conceivable being or there is more than one.

According to the Muslims' Hadith, there are 99 names of God. Nearly all of them begin with 'the', suggesting that the names are definite descriptions rather than names: The Almighty, The Irresistible, the All-Knowing, and on and on . . . actually for more than 99 names (if you count each form of divine address in the Koran).

According to Russell, there is no logically proper name of God. There is only a predicate 'god' as in 'The gods dwell on Mount Olympus'. Russell says that there are no gods, just as there are no ghosts. This grammatically negative claim is equivalent to the positive universal proposition that each thing is not a god (just as each thing is not a ghost). A god who was omnipotent, omniscient, and omnibenevolent would have made his existence obvious. For this god, absence of evidence is evidence of absence. This does not apply to self-concealing gods who use their powers to make the world appear godless.

Religious Education

'Agnostic' was coined in 1869 by Thomas Huxley, "Darwin's Bulldog," to denote the belief that human beings do not have enough evidence to ascertain whether God exists.

If 'possible' means logically possible, then 'God is possible' entails that God is actual by virtue of the ontological argument (recall Leibniz's interest in vindicating the ontological proof of God's existence by demonstrating that 'God exists' is not a contradiction). So Huxley needs to construe the possibility as whatever is compatible with what is known. And indeed, Huxley does talk as if the problem was that our ape brains are overmatched by the question. Huxley does not explain how we are a match for the higher level question of ascertaining which questions we are capable of answering and which questions we cannot possibly answer.

Knowing whether a name applies is easier than knowing whether a predicate applies. For instance, Russell was agnostic about whether Jesus existed but could easily envisage archaeological evidence that would have given him knowledge that Jesus existed. One can know a name has a bearer without knowing the nature of that bearer. Russell could not see how to resolve his agnosticism about whether there is a god. Most agnosticism is predicate agnosticism rather than name agnosticism.

If belief depends on evidence, then the absence of evidence of a god implies an absence of belief in a god. But if belief does not depend on evidence, then a theist could also be an agnostic. Indeed, some theologians claimed to be agnostics. They quote from the preface of Immanuel Kant's *Critique of Pure Reason*: "I had to deny knowledge in order to make room for faith" ([1781] 1965, Bxxx).

The distinction between proof and belief was also important for some atheists. They claimed that for propositions concerning existence, belief that x does not exist is the methodological default. Less is best.

Other philosophers deny there is any methodological asymmetry in questions of existence. They disbelieve there are ghosts only if there is evidence against ghosts. If there is no evidence for or against the existence of ghosts, then we should suspend judgment about whether ghosts exist. Impartial doubt rather than partisan doubt is the default.

According to Spinoza, our psychological default is belief rather than disbelief. We are as much prone to false starts as sprinters poised at a starting line.

In addition to agreeing with Spinoza's psychological thesis, Russell agreed with W. K. Clifford's normative thesis; we ought to suspend judgment in the absence of evidence. When there is evidence, we should proportion belief to that evidence. Russell thought it obvious that our frail constitutions make us fall short of this ideal rationality.

If we weary in our suppression of belief, latent conviction may spring forth. Under stress, many agnostics revert to childhood theism (though, Russell insists, not as many as aficionados of death-bed conversions claim).

Reversing a belief also requires effort. Instead of erasing the prior belief, disbelief complicates belief by adding a negation. As the atheist lies dying, his mental state simplifies. 'Theism' is the last word in 'atheism'.

Given this default theory of belief there should be a strong primacy effect for persuasion. The first word in religious affiliation tends to be the last word.

To prevent irrational imprinting, freethinkers condemn religious indoctrination as child abuse. Children should decide important matters for themselves after they are mature enough to weigh the evidence.

If parents are entitled to control the religious instruction of the children, does this extend to freethinkers opposed to religious instruction? And does this right persist after their deaths? In 1874, Bertrand Russell's mother and sister died of diphtheria. Sick with grief and just plain sick, Viscount Amberley amended his will. Explaining that he did not want his remaining children to be reared as Christians, the ailing widower designated two agnostic friends as guardians. Upon his death, however, the viscount's parents successfully petitioned the Court of Chancery to set aside this anti-Christian provision. Guardianship of the children passed to Lord John Russell, a former prime minister of Great Britain, and the pious Lady Russell.

Their views on religious training matched Samuel Taylor Coleridge's. He listened to a freethinker's objections to impressing religion on immature brains. Then, I showed him my garden, and told him it was my botanical garden. "How so?" said he; "it is covered with weeds." "Oh," I replied, "that is only because it has not yet come to its age of discretion and choice. The weeds, you see, have taken the liberty to grow, and I thought it unfair in me to prejudice the soil towards roses and strawberries" (2019, 310).

Childish Faith and Objectless Belief

Baby Bertrand Russell became a pious boy. Memories of his nebulous childhood faith made Russell receptive to William James's emotional theory of belief. Just as there are objectless emotions such as free-floating anxiety, there is free-floating belief. "Everyone knows the difference between imagining a thing and believing in its existence, between supposing a proposition and acquiescing in its truth . . . in its inner nature, belief, or the sense of reality, is a sort of feeling more allied to the emotions than to anything else" (James [1890] 1981, 2:283). Drugs that stimulate emotion, such as alcohol, intensify conviction at the expense of clarity.

Russell speculates that this "feeling of belief leads us to look for a content to which to attach it. Much of what passes for revelation or mystic insight probably comes in this way: the belief-feeling, in abnormal strength, attaches itself, more or less accidentally, to some content which we happen to think of at the appropriate moment" (1921, 252).

William James extended his emotional account to philosophies of oneness. When breathing nitrous oxide, William James was profoundly convinced. But of what? When the nitrous oxide was removed, the insights retreated to the tip of his tongue. The closest articulation he could find were those passages of Hegel reminiscent of Angel Falls in Venezuela—where the water evaporates before touching ground. These writings bridle against the law that everything is identical to itself and that no proposition is both true and false. The intoxicated James anticipates that his jottings will seem incoherent to those who filter reality through principles such as everything is identical to itself:

> Don't you see the difference, don't you see the identity? Constantly opposites united! The same me telling you to write and not to write! Extreme—extreme, extreme! Within the extensity that "extreme" contains is contained the "extreme" of intensity Something, and other than that thing! . . .
>
> By George, nothing but othing! That sounds like nonsense, but it's pure onsense! Thought much deeper than speech . . . ! Medical school; divinity school, school! SCHOOL! Oh my God, oh God; oh God! (James 1882, 207)

The oceanic nature of the experience lent support to the explanation that reality transcended verbal analysis.

Ultimately, James interpreted these Hegelian drug experiences as evidence against Hegel. James's self-experiments suggest that Hegel was naturally prone to the exaltation that James experiences only through nitrous oxide intoxication.

Ordinary people interpret the oceanic experience in the idiom of their religions. Hegel's education changed the vocabulary and became a substitute for religion. Russell interpreted his early Hegelianism as an intellectually fashionable substitute for religion.

Under Karl Marx's inversion of the Hegelian system, atheistic communism became a sublimated religion. Russell offers a dictionary to readers of the *Communist Manifesto* and *Das Kapital*:

Yahweh = dialectical materialism
the Messiah = Marx
the elect = the proletariat
the church = the Communist Party
the Second Coming = the revolution
Hell = punishment of the capitalists
The millennium = the communist commonwealth

The Soviets hoped Russell would graduate from Hegelianism to Marxism. But a visit to Moscow in 1920 frosted Russell's budding communism. His comrades heard the utilitarian hubbub of cars backfiring in the night. Russell heard execution squads.

Russell's drive to believe eventually became a drive to disbelieve. In politics, he defined himself oppositionally (anti-war, anti-segregation, anti-et cetera). His opposition to war, which only abated for World War II, resumed permanently in old age. In 1969, Russell organized international commissions of inquiry. Jean-Paul Sartre joined as well—though Sartre refused to consider allegations that communists committed war crimes.

Russell's Existential Crisis

Countersuggestibility is manifest in Arthur Schopenhauer's outspoken atheism. Bertrand Russell wanted to agree with Schopenhauer. However, Russell became puzzled as to what he would be agreeing.

John Stuart Mill's godson also became puzzled by disbelief in God—or anything else with a name. In classical logic, a name must have a bearer. If 'God' is a logically proper name, then 'God exists' is a tautology. Name atheism is contradictory.

Name agnosticism would also be self-contradictory. If 'God' is a logically proper name, then 'God might exist or might not exist' entails that God exists. Since tautologies are knowable and agnosticism entails that God is not knowable, name agnosticism is necessarily false.

If 'God' is not a name, then 'God does not exist' is as ungrammatical as 'San Ababís does not exist.' In short, 'God exists' cannot be false. It is either a tautology or meaningless.

For Russell, a name resembles a demonstrative term such as 'this' and 'that.' (Indeed, when Russell was a phenomenalist, he insisted 'this' and 'that' are the only logically proper names.) Saying 'That does not exist' is either trivially true or ill-defined. To be meaningful, the speaker's 'that' must match what he is pointing to. If there is nothing there, his pointing gesture failed to secure a topic to comment upon. Instead of being false, the demonstrative misfires.

Russell also wished to deny the existence of Satan. But what is the difference between 'God does not exist' and 'Satan does not exist'? If the meaning of a name is its bearer, then the absence of a bearer would preclude a difference in meaning between 'God' and 'Satan.'

The agnostic has a similar dilemma. He would like to confess that he does not know whether 'God does not exist' is true. But if he knows what he is referring to, he must know that God exists.

Even the theist has trouble. For he does not wish to affirm the existence of all gods.

Does God subsist?

Russell's first published views on negative existentials invoked a distinction between existence and subsistence—which he attributed to Alexius Meinong.[1] Zeus does not *exist* but he does *subsist*. Subsistence

[1] Meinong's distinction between existence and subsistence never existed. Further, Meinong was anticipated by Mill. Both treat ideas as backup referents. Gottlob Frege denies this is plausible for natural language. He recommends we adopt an artificial language in which some object, say the moon, is stipulated as the backup referent.

suffices to make 'Zeus does not exist' about Zeus. Meinong explored the realm of subsistence with the ambition of a conquistador.

Russell eventually became disenchanted by Meinong's approach. Philosophers should have the same robust sense of reality as Charles Darwin displayed when boys attempted to trick him with a specimen fused together from several insect parts. Pressed to identify the Frankenstein insect, Darwin taxonomized it as a humbug.

In "On Denoting" Russell presents a three-step program for analyzing negative existentials. First, show how that each indefinite description, such as 'a ghost', is a general term making no reference to any particular thing. I can infer there is a ghost from a house being haunted without being able to answer 'To which ghost are you referring?' The statement, 'It is false that a ghost exists', in his notation~(∃x)Gx, is equivalent to the universal statement 'Each thing is a nonghost', (x)~Gx.

Russell is here following Henri Bergson's advice to look for the positive statement behind the negative statement. Bergson's policy harmonizes with the sermon Johnny Mercer put to song in 1944 "Accentuate the Positive":

You've got to accentuate the positive
Eliminate the negative
And latch on to the affirmative
Don't mess with Mister In-Between

Russell's key move is to exploit the distinction between referring expressions and variables. A variable works like a blank. Filling in the blank makes complete sentences. When different entries must be added, the letters *x*, *y*, and *z* are reserved to make the selection restrictions: *x* is between *y* and *z*. The letters are just placeholders.

Russell's second step is to show how definite descriptions (the ghost) can be analyzed in terms of indefinite descriptions (a ghost): 'It is false that the ghost exists', ~(∃x)((Gx & (y)(Gy ⊃ y = x), is equivalent to the disjunction 'Either there are no ghosts or there are at least two ghosts': (x)~Gx v (∃y)(∃z)(Gy & Gz & ~(y = z)). Russell stresses the uniqueness implication of 'the'.

The final and most controversial step of Bertrand Russell's analysis is to unmask ordinary "names" as disguised definite descriptions. This extends the solution to negative existentials featuring names (such as 'Casper does not exist'), Russell denied that ordinary names are logically proper names. In formal logic, names behave just as John Stuart Mill contends. A name's sole meaning is its bearer. But ordinary names do not tag an object for future reidentification. They set up a profile that may net a single thing.

Specifically, Russell argued that ordinary names are disguised definite descriptions. If asked "Whom do you mean by 'Casper'?," you reply, "The friendly ghost." Russell could also appeal to descriptive uses of names: 'The Headless Horseman is no Casper'. If names were mere tags, then no content would be conveyed by such descriptions.

Now let us paraphrase the negative existential: 'Casper does not exist'. If 'Casper' were a logically proper name, then 'Casper does not exist' would indeed be untrue. For a logically proper name is only a label whose meaning is its bearer. No bearer, no meaning. No meaning, no truth.

However, 'Casper' means the friendly ghost. Hence 'Casper does not exist' is equivalent to 'The friendly ghost does not exist'. 'Casper does not exist' manages to be true because it is really a general description, rather than a remark about a particular thing. The statement really says that either there is more than one friendly ghost or there are no friendly ghosts: $(x){\sim}(Fx \,\&\, Gx) \vee (\exists y)(\exists z)((Fy \,\&\, Gy) \,\&\, (Fz \,\&\, Gz) \,\&\, {\sim}(y = z))$.

The application to atheism can be illustrated under the assumption that 'God' is a definite description such as 'the perfect being'. 'God does not exist' is then equivalent to the disjunctive assertion that either there is no being that is the perfect being or there are at least two such beings. A polytheist would be an atheist under this definition of 'God'. And indeed, the Christians in late antiquity applied 'polytheist atheist' to anyone who believed there was more than one god.

Arguing in Absentia

Bertrand Russell died in 1970, at 97. He had read his obituaries a half century earlier. In 1920, his double pneumonia had been exaggerated to death. Since Russell was lecturing in China, he had trouble correcting the rumors. Drama-hungry Japanese reporters wanted

Russell to provide living testimony that he was not dead. Fearing a relapse, the convalescing Russell declined. Disappointed, the Japanese reporters refused to correct reports of Russell's death. They hoped this would flush out Russell. Annoyed, Russell continued to dodge Japanese reporters even after recuperating enough for a trip to Japan. His secretary (and future wife) Dora Black interposed herself in all communication. Miss Black handed out placards: "Mr. Bertrand Russell, having died according to the Japanese press, is unable to give interviews to Japanese journalists" (Russell 1975, 144).

Though he no longer exists, Russell will continue to be the subject of debate and controversy. This includes controversy over the meaningfulness of the preceding sentence.

References

Allen, Joseph P. 1972. "Summary of Scientific Results." In NASA Manned Spacecraft Center, *Apollo 15 Preliminary Science Report*, 2-1–2-11. Washington, DC: National Aeronautics and Space Administration, Scientific and Technical Information Office.

Anscombe, Elizabeth. 1981. *From Parmenides to Wittgenstein*. Oxford: Blackwell.

Aristotle. 1941. *The Basic Works of Aristotle*. Edited by Richard McKeon. New York: Random House.

Augustine. 1872. *City of God*. Translated by M. Dods. Edinburgh: T. & T. Clark.

Augustine. 1948. "Concerning the Teacher." In *Basic Writings of Saint Augustine*, vol. 1, edited by Whitney J. Oates, 351–95. New York: Random House.

Augustine. 1953. "Of True Religion." In *Earlier Writings*, edited and translated by John H. S. Burdleigh, 225–283. Philadelphia: Westminster Press.

Augustine. 1963. *The Trinity*. In *Fathers of the Church*, vol. 45, translated by Stephen McKenna. Washington, DC: Catholic University of America Press.

Augustine. 1982. *The Literal Meaning of Genesis* vol. 1. Translated by John Hammond Taylor. Vol. 41 of *Ancient Christian Writers: The Works of the Fathers in Translation*. Edited by Johannes Quastern, Walter J. Burghardt, and Thomas Comerford Lawler. New York: Newman.

Augustine. 2010a. *Confessions*. Translated by V. J. Bourke. Washington, DC: Catholic University of America Press.

Augustine. 2010b. *On the Free Choice of the Will, On Grace and Free Choice, and Other Writings*. Edited and translated by Peter King. New York: Cambridge University Press.

Augustine. 2007. *Nicene and Post-Nicene Fathers First Series, St. Augustine: Gospel of John, First Epistle of John, Soliliques*. Translated by Schaff, Philip. New York: Cosimo Classics, (1888) 2007.

Austin, J. L. 1975. *How to Do Things with Words*. 2nd ed. Edited by M. Sbisà and J. O. Urmson. Oxford: Oxford University Press.

Bacon, Francis. 1859. *Apophthegms*. In *The Works of Francis Bacon*, edited by Basil Montagu. Philadelphia: Parry & McMillan.

Bergson, Henri. (1911) 1944. *Creative Evolution*. Translated by A. Mitchell. New York: Modern Library.

Bergson, Henri. (1932) 1977. *The Two Sources of Morality and Religion*. Translated by R. Ashley Audra and Cloudsley Brereton, with the assistance of W. Horsfall Carter. Notre Dame, IN: University of Notre Dame Press.

Bergson, Henri. (1946) 1992. *The Creative Mind*. Translated by Mabelle L. Andison. New York: Citadel Press.

Berkeley, George. 1948. *The Works of George Berkeley, Bishop of Cloyne*. Edited by A. A. Luce and T. E. Jessop. 9 vols. London: Thomas Nelson and Sons.

Bostrom, Nick. 2003. "Are We Living in a Computer Simulation?" *Philosophical Quarterly* 53 (211): 243–55.

Brewster, D. 1855. *Memoirs of the Life, Writings and Discoveries of Sir Isaac Newton*. Vol. 2. Edinburgh: Constable.

Bruner, Katherine Frost. 1942. "Of Psychological Writing: Being Some Valedictory Remarks on Style." *Journal of Abnormal and Social Psychology* 37 (1): 52–70.

Buddha. 1976. *The Dhammapada: The Sayings of the Buddha*. A new rendering by Thomas Byrom. New York: A. A. Knopf.

Camus, Albert. 1953. *The Myth of Sisyphus and Other Essays*. Translated by Justin O'Brien. New York: Alfred A. Knopf.

Casati, Roberto. 2018. "Shadows, Objects, and the Lexicon." In *Perceptual Ephemera*, edited by Thomas Crowther and Clare Mac Cumhaill, 154–71. New York: Oxford University Press.

Casati, Roberto, and Patrick Cavenagh. 2019. *The Visual World of Shadows*. Cambridge, MA: MIT Press.

Casati, Roberto, and Achille Varzi. 1994. *Holes and Other Superficialities*. Cambridge, MA: MIT Press.

Chalmers, David J. 2022. *Reality+: Virtual Worlds and the Problems of Philosophy*. New York: Norton.

Cicero.1923. *de Divinatione* translated by W.A. Falconer vol. XX. Cambridge, Massachusetts: Harvard University Press.

Clarke, Arthur C. 1969. "Beyond the Moon: No End." *Time*, July 18.

Coleridge, Samuel T. 2019. *The Collected Works of Samuel Taylor Coleridge*, Vol. 14: Table Talk, Part II. Princeton, New Jersey: Princeton University Press.

Confucius. 1871. *The Great Learning*. Translated by James Legge. London: N. Trübner.

Curd, Patricia. 2002. "The Metaphysics of Physics: Mixture and Separation in Empedocles and Anaxagoras." In *Presocratic Philosophy: Essays in Honour of Alexander Mourelatos*, edited by Victor Caston and Daniel W. Graham, 139–58. New York: Routledge.

Da Vinci, Leonardo. 1978. *The Codex Atlanticus of Leonardo Da Vinci : a Catalogue of Its Newly Restored Sheets*. Edited by Carlo Pedretti. New York: Johnson Reprint Corp.

Dante, Alighieri. 1982. *Dante's Cantos: Lines of Inferno*. Translated C. Sreechinth. Uintah Basin, Utah: UB Tech.

DeMorgan, Augustus. 1915. *A Budget of Paradoxes*. 2nd ed. Vol. 2. London: Open Court.

Diels, H., and W. Kranz. 1951. *Die Fragmente der Vorsokratiker*. 6th ed. Berlin: Weidmann.

Dieter, K. C., B. Hu, D. C. Knill, R. Blake, and D. Tadin. 2014. "Kinesthesis Can Make an Invisible Hand Visible." *Psychological Science* 25: 66–75.

Diogenes Laertius. 1925. *Lives of Eminent Philosophers*. Translated by R. D. Hicks. 2 vols. Cambridge, MA: Harvard University Press.

Donnellan, Keith. 1974. "Speaking of Nothing." *Philosophical Review* 83: 3–31.

Duchamp, Marcel. 1919. L.H.O.O.Q published in 391, n. 12, March 1920.

Einstein, Albert. (1918) 2002. "Motives for Research." Doc. 7 in *The Collected Papers of Albert Einstein*, vol. 7, edited by John Stachel et al., 41–44. Princeton, NJ: Princeton University Press.

Epictetus. 1916. *The Discourses and Manual*. Translated by P. E. Matheson. Oxford: Clarendon Press.

Fadiman, Clifton. 1985. *The Little, Brown Book of Anecdotes*. Boston: Little, Brown.

Fridugsis of Tours. 1995. *On the Being of Nothing and Shadows*. Translated by Paul Vincent Spade, downloaded from https://scholarworks.iu.edu.

Gale, Richard M. 1976. *Negation and Non-Being*. London: William Clowes & Sons.

Grant, Edward. 1981. *Much Ado about Nothing: Theories of Space and Vacuum from the Middle Ages to the Scientific Revolution*. Cambridge: Cambridge University Press.

Guericke, Otto von. 1672. *Experimenta Nova (ut vocantur) Magdeburgica de Vacuo Spatio*. Amsterdam: Joannes Jansson à Waesberg.

Gray, Jeffrey Allan. 1987. *The Psychology of Fear and Stress*. Cambridge: Cambridge University Press.

Hall, Katherine. 2019. "Did Alexander the Great Die from Guillain-Barré Syndrome?" *Ancient History Bulletin* 32: 106–128.

Harman, Gilbert. 1963. "Generative Grammars without Transformational Rules: A Defense of Phase Structure." *Language* 39: 597–616.

Harman, Gilbert. 1988. "The Simplest Hypothesis." *Critica* 20 (59): 23–42.

Hawking, Stephen, and Leonard Mlodinow. 2010. *The Grand Design*. New York: Bantam Books.

Hegel, G. W. F. 1929. *Science of Logic*. Translated by W. H. Johnson and G. L. Slaughter, Vol. 1. London: Allen and Unwin.

Homer. 1911. *The Odyssey*. Translated by H. B. Cotterill. London: George G. Harrap.

Herodotus. 1920. *The Histories*. Translated by A. D. Godley. Cambridge, MA: Harvard University Press.

Huemer, Michael. 2021. "Existence Is Evidence of Immortality." *Noûs* 55 (1): 128–51.

James, William. 1882. "On Some Hegelisms." *Mind* 7: 186–208.

James, William. (1890) 1981. *The Principles of Psychology*. 2 vols. Cambridge, MA: Harvard University Press.

Judovitz, Dalia. 1987. "Rendez-vous with Marcel Duchamp: Given." *Dada/Surrealism* 16, 184–202. Iowa City, Iowa: University of Iowa.

Kant, Immanuel. (1781) 1965. *Critique of Pure Reason*. Translated by N. K. Smith. London: Macmillan.

Kant, Immanuel. (1790) 1911. *Critique of Judgment*. Translated by James Creed Meredith. 2 vols. Oxford: Clarendon Press.

Kant, Immanuel. (1785) 1996. *The Metaphysics of Morals*. Translated by Mary Gregor. Cambridge, UK: Cambridge University Press.

Kripke, Saul. 1980. *Naming and Necessity*. Oxford: Basil Blackwell.

Lao-tzu. 1974. *Tao Te Ching*. Translated by Gia-fu Feng and Jane English. London: Wildwood House.

Leder, D., Hermann, R., Hüls, M., Russo, G., Hoelzmann, P., Nielbock, R., Böhner, U., Lehmann, J., Meier, M., Schwalb, A., Tröller-Reimer, A., Koddenberg, T., and Terberger, T. A. "51,000-year-old engraved bone reveals Neanderthals' capacity for symbolic behaviour." *Nature Ecology and Evolution*. 2021 Jul 5. doi: 10.1038/s41559-021-01487-z. Epub ahead of print. Erratum in: Nat Ecol Evol. 2021 Jul 20: PMID: 34226702.

Leibniz, Gottfried Wilhelm. (1686) 1966. "General Inquiries about the Analysis of Concepts and of Truths." In *Logical Papers*, edited by G. H. R. Parkinson, 47–87. Oxford: Clarendon Press.

Leibniz, Gottfried Wilhelm. 1703. *Die mathematische schriften von Gottfried Wilhelm Leibniz*, vol. VII C. I, edited by Gerhardt 1859: 223–227. Translated from the French by Lloyd Strickland. Berlin: A. Ascher.

Leibniz, Gottfried Wilhelm. 1966. *Logical Papers*. Edited by G. H. R. Parkinson. Oxford: Clarendon Press.

Leibniz, Gottfried Wilhelm. 1996. "On the Ultimate Origination of Things." In Leibniz, *New Essays on Human Understanding*, edited by Roger Ariew. Cambridge: Cambridge University Press.

Leslie, Alan. 1994. "Pretending and Believing: Issues in the Theory of ToMM." *Cognition* 50: 211–38.

Legge, James. 1891. *The Writings of Chuang Tzu*. Oxford: Oxford University Press.

Lewis, David, and Stephanie R. Lewis. 1970. "Holes." *Australasian Journal of Philosophy* 48: 206–12.

Lewis, Katherine J. 2000. *The Cult of St Katherine of Alexandria in Late Medieval England*. Woodbridge: Boydell.

Liddell, Henry George, and Robert Scott, rev. Henry Stuart Jones. 1968. *A Greek-English Lexicon*. Oxford: Oxford University Press.

Liu, JeeLoo. 2014. "Was There Something in Nothingness? The Debate on the Primordial State between Daoism and Neo-Confucianism." In *Nothingness in Asian Philosophy*, edited by JeeLoo Liu and Douglas Berger, 181–96. London: Routledge.

Locke, John. (1690) 1975. *An Essay Concerning Human Understanding*. edited by P. H. Nidditch. Oxford: Clarendon Press.

Longfellow, Henry Wadsworth. 1922. *The Complete Poetical Works of Longfellow*. Edited by Horace Elisha Scudder. Boston: Houghton Mifflin.

Lucian. 1901. "A Slip of the Tongue in Greeting." In *Dialogues and Stories from Lucian of Samosata*, translated by Winthrop Dudley Sheldon. Philadelphia: Drexel Biddle.

Lucretius. 1916. *De Rerum Natura*. Translated by William Ellery Leonard. New York: E. P. Dutton.

Lundgren, Erick J., et. al. 2021. "Equids Engineer Desert Water Availability." Science 30: 491–495.

Maimonides, Moses. 1904. *The Guide for the Perplexed*. Translated by M. Friedlander. London: George Routledge and Sons.

Mandel, Oscar. 1964. *Chi Po and the Sorcerer*. United Kingdom: C.E. Tuttle.

Merton, Thomas. 1969. *The Way of Chuang Tzu*. New York: New Directions.

Meinong, Alexius. 1904. "Über Gegenstandstheorie." Translated as 'The Theory of Objects.' In *Realism and the Background of Phenomenology*, edited by R. M. Chisholm Glencoe, 76–117. IL: Free Press, 1960; reprint: Atascadero, CA: Ridgeview, 1981.

Michotte, Albert. 1946. *The Perception of Causality*. Translated by T. R. Miles and E. Miles. London: Methuen, 1962.

Mill, John Stuart. (1882) 1979. *An Examination of Sir William Hamilton's Philosophy*. Edited by J. Robeson. Toronto: University of Toronto Press.

Millán, Elizabeth. 2017. "Aesthetic Humanism: Poetry's Role in the Work of Friedrich Schlegel and Schopenhauer." In *The Palgrave Schopenhauer Handbook*, edited by S. Shapshay, 179–96. London: Palgrave Macmillan.

Moore, G. E. 1903. "The Refutation of Idealism." *Mind*, new series, 12 (48): 433–53.

Moore, T., Bartholomew, B., Linsley, J. L., Dowden, W. S. 1983. *The Journal of Thomas Moore*. United Kingdom: University of Delaware Press.

Mumford, Stephen. 2021. *Nothing Really Matters*. Oxford: Oxford University Press.

Newton, Isaac. 2004. *Philosophical Writings*. Edited by A. Janiak. Cambridge: Cambridge University Press.

Nietzsche, Friedrich. (1889) 1998. *Twilight of the Idols*. Translated by R. J. Hollingdale. Oxford: Oxford University Press.

Odlyzko, Andrew. 2019. "Newton's financial misadventures in the South Sea Bubble." *The Royal Society Journal of the History of Science* 7329–7359.

Parfit, Derek. 1984. *Reasons and Persons*. Oxford: Oxford University Press.

Parsons, Terence. 1980. *Nonexistent Objects*. New Haven: Yale University Press.

Pascal, Blaise. (1669) 1910. *Pensées*. Translated by W. F. Trotter. New York: P. F. Collier & Sons.

Plato. 1892. *The Dialogues of Plato translated into English with Analyses and Introductions* by B. Jowett, M.A. in Five Vol. 3rd edition revised and corrected. Oxford University Press.

Plotinus. 1991. *The Enneads*. Translated by Stephen MacKenna, abridged and edited by John Dillon. London: Penguin.

Plutarch. 1914. *Plutarch's Lives and Writings*. Edited by A. H. Clough and William W. Goodwin. London: Simpkin, Hamilton, Kent.

Priest, Graham. 2016. *Towards Non-Being*. Oxford: Oxford University Press.

Proops, Ian. 2001. "Russell on Substitutivity and the Abandonment of Propositions." *Philosophical Review* 120(2): 151–205.

Rée, J. 1999. *I See a Voice*. London: HarperCollins.

Rekdal, Ole Bjørn. 2014. "Monuments to Academic Carelessness: The Self-Fulfilling Prophecy of Katherine Frost Bruner." Science, *Technology, and Human Values* 39 (5): 744–58.

Robinson, Jenefer. 1995. "Startle." *Journal of Philosophy*: 92(2): 53–74.

Russell, Bertrand. 1903. *The Principles of Mathematics*. Cambridge: Cambridge University Press.

Russell, Bertrand. 1905. "On Denoting." *Mind* 14: 479–93.

Russell, Bertrand. 1919. *Introduction to Mathematical Philosophy*. London: George Allen and Unwin.

Russell, Bertrand. 1921. *The Analysis of Mind*. London: George Allen and Unwin.

Russell, Bertrand. 1940. *An Inquiry into Meaning and Truth*. London: George Allen and Unwin.

Russell, Bertrand. 1948. *Human Knowledge: Its Scope and Limits*. London: George Allen and Unwin.

Russell, Bertrand. 1959. *My Philosophical Development*. London: George Allen and Unwin.

Russell, Bertrand. 1968. *Autobiography*. London: George Allen and Unwin.

Russell, Bertrand. 1985. *The Philosophy of Logical Atomism*. La Salle, IL: Open Court.

Russell, Bertrand, and Alfred North Whitehead. 1910–13. *Principia Mathematica*. Cambridge: Cambridge University Press.

Russell, D. W. B. R. 1975. *My Quest for Liberty and Love*. United States: Putnam.

Sainsbury, Robert. 1973. "Discrimination Learning Utilizing Positive or Negative Cues." *Canadian Journal of Psychology* 27(1): 46–57.

Sartre, Jean Paul. 1964. *Words*. Translated by Irene Clephone. London: Hamish Hamilton.

Sartre, Jean Paul. 1969. *Being and Nothingness*. Translated by Hazel E. Barnes. New York: Washington Square Press.

Sartre, Jean Paul. 2004. *Imaginary*. Translated by Jonathan Webber. New York: Routledge.

Schopenhauer, Arthur. 1819. *The World as Will and Representation*. Translated by E. F. J. Payne. 2 vols. Indian Hills, CO: Falcon Wing Press, 1958.

Schopenhauer, Arthur. 1851. *Parerga and Paralipomena: Short Philosophical Essays*. Translated by E. F. J. Payne. (1974). Oxford: Oxford University Press.

Schopenhauer, Arthur. (1890) 2007. *Counsels and Maxims*. New York: Cosimo.

Schopenhauer, Arthur. 1912. *Counsels and Maxims: Being the Second Part of Arthur Schopenhauer's Aphorismen Zur Lebensweisheit*. Translated by T. Bailey Saunders, M.A. United Kingdom: S. Sonnenschein & Company.

Sedley, D. N. 1998. *Lucretius and the Transformation of Greek Wisdom*. Cambridge University Press.

Sidgwick, Henry. 1874. *The Methods of Ethics*. London: Macmillan.

Siderits, Mark. 2021. *Buddhism as Philosophy*. Hackett.

Sorensen, Roy. 2007. "The Vanishing Point: Modeling the Self as an Absence." *The Monist* 90(3): 432–456.

Sorensen, Roy. 2008. *Seeing Dark Things: The Philosophy of Shadows*. New York: Oxford University Press.

Sorensen, Roy. 2015. "Perceiving Nothings." In *Oxford Handbook of Perception*, edited by Mohan Matthen. Oxford University Press.

Strauss, Leo. 1952. *Persecution and the Art of Writing*. Chicago: University of Chicago Press.

Sze, Mai-mai, and Michael J. Hiscox. 2015. *The Mustard Seed Garden Manual of Painting: Chieh Tzu Yuan Hua Chuan, 1679–1701*. Princeton, NJ: Princeton University Press.

Taylor, C. C. W. 1999. *The Atomists: Leucippus and Democritus. Fragments, a Text and Translation with Commentary*. Toronto: University of Toronto Press.

Taylor, Richard. 1952. "Negative Things." *Journal of Philosophy* 49: 433–49.

Ulam, Stanislav. 1976. *Adventures of a Mathematician*. Berkeley: University of California Press.

Warren, J. 2004. *Facing Death: Epicurus and His Critics*. New York: Oxford University Press.

Whitehead, Alfred North. (1929) 1978. *Process and Reality*. New York: Free Press.

Index

For the benefit of digital users, indexed terms that span two pages (e.g., 52–53) may, on occasion, appear on only one of those pages.

Figures are indicated by *f* following the page number